Lecture Notes in Computer Science 8141

Commenced Publication in 1973
Founding and Former Series Editors:
Gerhard Goos, Juris Hartmanis, and Jan van Leeuwen

T0212947

David Soloveichik   Bernard Yurke (Eds.)

DNA Computing and Molecular Programming

19th International Conference, DNA 19
Tempe, AZ, USA, September 22-27, 2013
Proceedings

 Springer

Volume Editors

David Soloveichik
University of California, San Francisco
Center for Systems and Synthetic Biology
1700 4th Street
San Francisco, CA 94158, USA
E-mail: david.soloveichik@ucsf.edu

Bernard Yurke
Boise State University
Materials Science and Engineering Department
1910 University Drive
Boise, ID 83725, USA
E-mail: bernardyurke@boisestate.edu

ISSN 0302-9743   e-ISSN 1611-3349
ISBN 978-3-319-01927-7   e-ISBN 978-3-319-01928-4
DOI 10.1007/978-3-319-01928-4
Springer Cham Heidelberg New York Dordrecht London

Library of Congress Control Number: 2013946048

CR Subject Classification (1998): F.1, F.2.2, J.3, E.1, I.2, G.2, F.4

LNCS Sublibrary: SL 1 – Theoretical Computer Science and General Issues

Typesetting: Camera-ready by author, data conversion by Scientific Publishing Services, Chennai, India

Printed on acid-free paper

Springer is part of Springer Science+Business Media (www.springer.com)

Preface

This volume contains the refereed proceedings of DNA19: the 19th International Conference on DNA Computing and Molecular Programming, held September 22–27, 2013, at Arizona State University, Arizona, USA.

Research in DNA computing and molecular programming draws together many disciplines (including mathematics, computer science, physics, chemistry, materials science, and biology) to address the analysis, design, and synthesis of information-based molecular systems. This annual meeting is the premier forum where scientists with diverse backgrounds come together with the common purpose of applying principles and tools of computer science, physics, chemistry and mathematics to advance molecular-scale computation and nanoengineering. Continuing this tradition, the 19th International Conference on DNA Computing and Molecular Programming (DNA19), organized under the auspices of the International Society for Nanoscale Science, Computation, and Engineering (ISNSCE), focused on important recent experimental and theoretical results.

This year the Program Committee received 29 full paper submissions and 30 one-page abstracts. The Committee selected 14 full papers for oral presentation and inclusion in these proceedings, and 14 abstracts were selected for oral presentation only. Many of the remaining submissions were selected for poster presentation, along with posters chosen from 51 additional poster-only submissions. The topics were well-balanced between theoretical and experimental work, with submissions from 18 countries, reflecting the diversity of the community.

The scientific program also included six invited speakers: Alessandra Carbone (Pierre and Marie Curie University), Hendrik Dietz (Technical University of Munich), Eric Goles (Adolfo Ibáñez University), Chengde Mao (Purdue University), Lulu Qian (California Institute of Technology), and Yannick Rondelez (University of Tokyo).

Following the conference, a one-day workshop, Nanoday 2013, was held featuring current topics in nanotechnology. The speakers included Nadrian Seeman (New York University), Friedrich Simmel (Technical University Munich), John Spence (Arizona State University), Kurt Gothelf (Aarhus University), Tim Liedl (Ludwig Maximilians University of Munich), William Shih (Harvard Medical School), John Chaput (Arizona State University), Paul Steinhardt (Princeton University), Hanadi Sleiman (McGill University), Peng Yin (Harvard Medical School), and Dongsheng Liu (Tsinghua University).

The editors would like to thank the members of the Program Committee and the external referees for their hard work in reviewing submissions and offering constructive suggestions to the authors. They also thank the Organizing and Steering Committees, and particularly Hao Yan and Natasha Jonoska, the respective Committee Chairs, for their invaluable help and advice. Finally, the editors would like to thank all the sponsors, authors, and attendees for supporting the DNA computing and molecular programming community, and helping to make the conference a success.

July 2013   David Soloveichik
                                                                         Bernard Yurke

Organization

DNA19 was organized by Arizona State University in cooperation with the International Society for Nanoscale Science, Computation, and Engineering (ISNSCE).

Steering Committee

Natasha Jonoska (Chair)	University of South Florida, USA
Luca Cardelli	Microsoft Research Cambridge, UK
Anne Condon	University of British Columbia, Canada
Masami Hagiya	University of Tokyo, Japan
Lila Kari	University of Western Ontario, Canada
Satoshi Kobayashi	University of Electro-Communication, Japan
Chengde Mao	Purdue University, USA
Satoshi Murata	Tohoku University, Japan
John Reif	Duke University, USA
Grzegorz Rozenberg	University of Leiden, The Netherlands
Nadrian Seeman	New York University, USA
Friedrich Simmel	Technische Universität München, Germany
Andrew J. Turberfield	Oxford University, UK
Hao Yan	Arizona State University, USA
Erik Winfree	California Institute of Technology, USA

Organizing Committee

Hao Yan (Chair)	Arizona State University, USA
Jeanette Nangreave	Arizona State University, USA
Yan Liu	Arizona State University, USA

Program Committee

David Soloveichik (Co-chair)	University of California, San Francisco, USA
Bernard Yurke (Co-chair)	Boise State University, USA
Robert Brijder	Hasselt University, Belgium
Luca Cardelli	Microsoft Research Cambridge, UK
Andrew Ellington	University of Texas at Austin, USA
Elisa Franco	University of California, Riverside, USA
Kurt Gothelf	Aarhus University, Denmark
Natasha Jonoska	University of South Florida, USA
Lila Kari	University of Western Ontario, Canada
Satoshi Kobayashi	University of Electro-Communications, Japan

Dongsheng Liu	Tsinghua University, China
Chengde Mao	Purdue University, USA
Yongli Mi	Hong Kong University of Science and Technology, China
Satoshi Murata	Tohoku University, Japan
Jennifer E. Padilla	New York University, USA
Matthew J. Patitz	University of Arkansas, USA
Andrew Phillips	Microsoft Research Cambridge, UK
John Reif	Duke University, USA
Alfonso Rodríguez-Patón	Universidad Politécnica de Madrid, Spain
Yannick Rondelez	University of Tokyo, Japan
Robert Schweller	University of Texas-Pan American, USA
Shinnosuke Seki	Aalto University, Finland
Friedrich Simmel	Technical University of Munich, Germany
Darko Stefanovic	University of New Mexico, USA
Chris Thachuk	University of British Columbia, Canada
Andrew J. Turberfield	University of Oxford, UK
Erik Winfree	California Institute of Technology, USA
Damien Woods	California Institute of Technology, USA
Hao Yan	Arizona State University, USA
Peng Yin	Harvard Medical School, USA
Byoung-Tak Zhang	Seoul National University, South Korea

External Reviewers

Hieu Bui	Niall Murphy
Xi Chen	Cameron Myhrvold
Mingjie Dai	Mark Olah
Sudhanshu Garg	Andrei Paun
Nikhil Gopalkrishnan	Trent A. Rogers
Alireza Goudarzi	John Sadowski
Martín Gutiérrez	Iñaki Sainz de Murieta
Jacob Hendricks	Amir Simjour
Kojiro Higuchi	Tianqi Song
Ralf Jungmann	Petr Sosik
Korbinian Kapsner	Milan Stojanovic
Steffen Kopecki	Scott Summers
Matthew R. Lakin	Wei Sun
Terran Lane	Maximilian Weitz
Reem Mokhtar	Justin Werfel

Sponsoring Institutions

The National Science Foundation
The US Army Research Office
The US Office of Naval Research
The Office of Knowledge Enterprise Development, the Biodesign Institute, and the Department of Chemistry and Biochemistry at Arizona State University

Table of Contents

Extending DNA-Sticker Arithmetic to Arbitrary Size Using Staples

Mark G. Arnold

XLNS Research

Abstract. The simple-sticker model uses robotic processing of DNA strands contained in a fixed number of tubes to implement massively-parallel processing of bit strings. The bits whose value are '1' are recorded by short DNA "stickers" that hybridize at specific places on the strand. Other DNA models, like folded origami, use "staples" that hybridize to disjoint portions of a single strand. This paper proposes an extended-sticker paradigm that uses staples to hybridize to contiguous portions of two substrands, forming virtual strands. The problem of redundant bits is solved by blotting out old values. As an example of the novel extended-sticker paradigm, a log-time summation algorithm outperforms (with an ideal implementation) any electronic supercomputer conceivable in the near future for large data sets. JavaScript and CUDA simulations validate the theoretical operation of the proposed algorithm.

Keywords: DNA arithmetic, sticker system, DNA staples, addition.

1 Introduction

Roweis et al. [19] introduced the *sticker system*, for processing information encoded using carefully-designed species of single-stranded DNA contained in a small set of tubes. Using an ordinary embedded microprocessor, the molecular-level steps for the user's algorithm are performed by macro-scale robotic processing (moving, warming, cooling, and filtering) of water in particular tubes carrying DNA strands. Although in recent years DNA-sticker computation has largely been neglected by the molecular computing community that it helped establish, there appears to be some renewed interest in sticker-like computation [3,4]. Stickers were suggested as having the yet-to-be-realized parallel processing power to outperform electronic supercomputers (assuming the application is large and errors could be managed); newer non-sticker models of DNA computation avoid such comparisons, focusing on different paradigms, like nano-assembly. This paper proposes novel extensions to the sticker model that enhance its ability to solve larger problems. Although these ideas are specific to stickers, they may be worth considering in the broader context of molecular computing because of theoretical issues (e.g., sharing information between molecules) and implementation challenges (e.g., scaling up molecular computation to realistic applications) this paper considers.

Like most other DNA-computation models, the sticker system exploits the proclivity of a single-strand of DNA to hybridize with another single strand

D. Soloveichik and B. Yurke (Eds.): DNA 2013, LNCS 8141, pp. 1–15, 2013.

when the two strands are perfect Watson-Crick complements of each other. The literature suggests it is possible to design artificial [11] or discover naturally-occurring [18] sequences of DNA that hybridize with high-probability only at the sites expected in the user's high-level model.

The sticker system, as originally proposed [19], is one of the simplest models of DNA computation. At the user's level, it can be conceptualized without reference to the underlying biochemical implementation as processing k-bit strings in a finite set of containers (i.e., tubes). This model is isomorphic to punch-card data processing, which was widely used for automated business applications (e.g., radix-sort) between 1900 and 1960.

As envisioned in [19], the "address" of each of the k bits is represented by a distinct pattern of $m >> \log_4(k)$ nucleotides. The choice of m must be much larger than the information-theoretical bound to avoid unwanted hybridization of strands and stickers in the wrong places. The choice of the constants k and m determines the number and complexity of the species of DNA used to implement Roweis' simple-sticker system [19]. This simple-sticker model is capable of universal computation assuming k is large enough. Unfortunately, many problems need k larger than is practical (given the number of DNA species that can be managed simultaneously). The novel approach proposed here is an extended-sticker system, which allows solution of unbounded-size problems using a number of DNA species similar to a fixed-k simple-sticker system.

Unlike Adelman's complicated multi-tube method [1] using species of DNA created for a particular instance of a problem (i.e., Hamiltonian path), the simple-sticker system avoids using enzymes and permits recycling. Unlike single-tube DNA origami [18], where a long DNA strand is folded and held together by many species of *staples*, which consist of two ($\approx m$-sized) regions that are complementary to disjoint portions of the long strand, each simple sticker consists of one region that is complementary to a contiguous region of the long strand.

Unlike single-tube strand-displacement systems [17] that implement digital-circuit cascades autonomously, sticker systems need an electronic controller to initiate robotic processing between tubes. Strand-displacement systems involve toehold regions that define reversible hybridization reactions, unlike the simple-sticker system, where each sticker defines an essentially irreversible hybridization reaction (unless temperature or pH are changed by the controller).

Early simple-sticker applications focused on NP-complete algorithms [19,12]. Biochemical computation is less reliable than electronic computation. Compensating for this unreliability with redundancy [7] is now considered as too expensive for NP-complete problems[15]. Recently, Arnold [3] proposed biochemical unreliability can be useful in a different paradigm (Monte-Carlo arithmetic [16]).

There has been renewed interest in the simple-sticker paradigm for computer-arithmetic algorithms [22,10,2]. For example, Chang et al. [5] uses simple-sticker arithmetic for RSA cryptanalysis. Arnold [2,3] describes a new 50%-more-reliable-sticker algorithm for (tube-encoded) integer addition that does not need stickers to record carries, and uses half the time and tubes, and suggests Logarithmic Number Systems (LNS) for real multiplication, division and square root. The novel extended-sticker approach proposed in this paper is compatible with all sticker-arithmetic methods in the literature, but yields the greatest performance with the tube-encoded methods [2].

2 Simple Sticker System

The simple-sticker system [19,12] consists of four types of DNA species. First, there is one kind of a long (mk-sized) single-stranded DNA molecule, which is referred to simply as the *strand* (since there is only one kind ever used). Second, there are k species of *stickers*, which are short (m-sized) complementary DNA molecules which hybridize only at one position in the long strand and never hybridize with each other. Whether a sticker is present or not at an arbitrary bit position in a particular strand determines whether that bit is treated as a '1' or as a '0'. Injecting a higher concentration of a particular type of sticker molecule will cause the associated bit to become '1' in essentially every strand. Third, there are k species of probes, with a subset of the DNA sequences (of the associated k stickers) covalently bonded to a substituent that can be macroscopically manipulated (e.g., magnetic bead) to separate the probe (and the strand it might hybridize) without disturbing other strands or unhybridized stickers. This allows strands to be separated into two different tubes depending on the value of the bit at the associated position. If a sticker is there, the probe cannot stick. If a sticker is not there, the strand will be transferred to a different tube along with the probe. The DNA sequence of the probes should probably be shorter than m, allowing the probe to be melted off the strand without disturbing the stickers on the strand in the new tube. Fourth, k anti-probes are used at the end of a problem after melting all stickers off the strands. Anti-probes allow for sorting and recycling of the k species of stickers for use in the next problem in the simple-sticker system. Like an operating system's privileged instructions, anti-probes are used only by the system and are not available to the user (e.g., combining probes and anti-probes in the same tube would incapacitate the system).

In addition to the n_t user tubes visible to the algorithm, an additional $3k$ system tubes hold the stickers, probes and anti-probes that implement individual sticker operations. The fundamental $O(1)$ user-level operations in the simple-sticker model [19,12] operate on bit positions $0 \le i < k$ within user tubes $0 \le t, t_0, t_1, t_2 < n_t$:

- separate(t_1, t_0, i) strands of tube t_0 that have a one at position i into tube t_1 and leave those that have a zero at that position in tube t_0.
- set(t, i): bit position i equal to 1 for every strand of tube t.
- discard(t): Recycle the contents of tube t back to the system tubes.
- combine(t_1, t_0): Transfer all of tube t_0 into tube t_1.
- split(t_1, t_0): Transfer half of tube t_0 into tube t_1.

Another operation, clear(t, i), which makes bit position i equal to 0, is often provided, but its biochemical implementation is considered problematic [12]. A more general form separate3(t_1, t_0, t_2, i) is also provided that transfers each strand from t_2 into t_1 or t_0 based on bit i. separate(t_1, t_0, i) means the same as separate3(t_1, t_0, t_0, i). These operations are implemented in JavaScript and CUDA simulators [2,3,21] which represent each sticker bit abstractly as an actual bit in the computer and do not consider lower-level DNA encoding. The JavaScript simulator uses that language's interpreter to evaluate (e.g., eval()) high-level sticker code supplied by a user on its web page; the much faster CUDA simulator requires the user to compile code (with nvcc) that calls on macros for the sticker operations. The CUDA code exploits the multi-core nature of GPUs to run each sticker as a separate thread, allowing for parallelism up to the amount of GPU hardware available. Two versions of the CUDA code exist: one which (like the JavaScript) assumes flawless operation; the other allows a user-supplied probability of misclassification (as in [7]). The former allows for simulation of larger systems because the latter must implement a pseudo-random number generator in each thread. These non-random simulators (that assume flawless sticker operation) were the starting points for the the extended-sticker simulators used later in this paper to validate the theoretical operation of the novel ideas presented here.

3   Novel Extended Sticker System

This paper proposes three novel extensions to the simple sticker system which allow unbounded-size data to be processed without adding significant complexity beyond that of the simple-sticker system. First, this paper proposes weak stickers that can be bulk erased without disturbing other stickers. Second, this paper suggests instead of putting all data on a single strand, with bits numbered consecutively from 0 to $k-1$, the data be distributed on several (n_s) substrands, with each substrand holding disjoint subsets of those bits. Third, this paper considers using staples to concatenate such substrands under programmed control. With these features, three new operations are available to the sticker programmer:

- clearweak(t) Remove all weak stickers from all strands in tube t.
- setweak(t, i): Make bit i equal to a weak 1 for every strand of tube t.
- staple(t_1, i, j, t_0): Randomly staple any substrand in tube t_1 whose bit at position i is zero to any substrand from tube t_0 whose bit at position j is zero. When finished, fully-stapled and completely-unstapled substrands from both tubes reside in t_1 (as if a combine had occured); half-stapled exceptions (if any) are placed in tube t_0.

Let's discuss each of these three novel features in more detail. First, weak stickers are simply stickers that are shorter than ordinary strong stickers, and therefore will melt away from the strand at lower temperature than the strong stickers. The mechanism is similar to how probes work. By itself, clearweak does not overcome the lack of a general clear. To understand why, consider a simple iterative algorithm, $a_{i+1} = f(a_i)$, that attempts to reuse an a field in a strand. Since any bit in the result might depend on any bit in the argument, there needs to be at least one other similar-sized field, b, that will hold a copy of a.

If we clear all bits of b at the start using a `clearweak`, we may then implement the assignment b=a using only `separate`s and `set`s. The problem is we must also clear the previous value held in a before we can compute a=f(b). If we attempted to use `clearweak` again here, we would lose the information needed to compute f(b). The different reason for `clearweak` here will be illustrated shortly. (Algorithms like $a_{i+1} = f(a_i)$ can be solved using techniques similar to those described below.)

Second, using substrands allows data to be processed independently (possibly in different tubes) prior to joining together into a data structure.

Third, staples are the key to making substrands join together. The implementation of $\texttt{staple}(t_1, i, j, t_0)$ shares aspects similar to combine and set. Like set, a larger concentration of short single-stranded DNA will be injected into t_1 and then filtered out. (The main distinction is these short molecules are $\approx 2m$-sized staples rather than $\approx m$-sized stickers.) Like combine, substrands from t_0 will be moved to t_1. (Unlike combine, exceptions are moved to t_0 when there are an excess of substrands in t_1 which causes some of them to become half-stapled.) The staple operation includes both combining, injecting and exception-filtering aspects as one user-level operation since it is necessary to return the excess staples to the system tubes prior to combining and there is no user-level command to do this because it involves anti-probes. If the unhybridized staples were not removed in the middle of the staple operation, some substrands from both tubes would end up half stapled, unable to connect to substrands from the opposite tube. As mentioned previously, the other way half-stapling can occur is when there are too many substrands in tube t_1. Again, the filtering of these exceptions to tube t_0 is included as part of the intrinsic staple operation because it involves anti-probes.

Because only a small number of staple and substrand types are provided, repeated stapling will result in a *virtual strand* with redundant copies of the same bit that do not necessarily agree in value. The nature of the separate operation (a probe may attach to any part of the virtual strand that does not have a sticker for this redundantly-occurring bit address), the bit value tested will be the logical conjunction of those redundant bits. The novel solution proposed here is to "blot" old values of these fields with ones, so that the result of the separate operation will be based only on the most recent value. In effect, this overcomes the lack of a general clear operation, at the cost of keeping "waste" product as part of the virtual substrand. This waste does not interfere with the theoretical operation of the system, but may impede practical implementation.

Fig. 1 a) shows the distinct single-stranded molecules in the novel extended-sticker/staple system with $k = 6$, $m = 7$ and number of substrands, $n_s = 2$. There are $k = 6$ stickers (solid-line backbones above the bases), four weak stickers (dashed-line backbones), $n_s = 2$ staples (longer backbones above the bases) and $n_s = 2$ substrands (even longer backbones below the bases). (In many algorithms—like the one presented later—positions that have a weak sticker (5, 3, 2, 0 in Fig. 1) may not also need a strong sticker.) The bit position on substrands can be determined by counting As; the bit position of staples or stickers

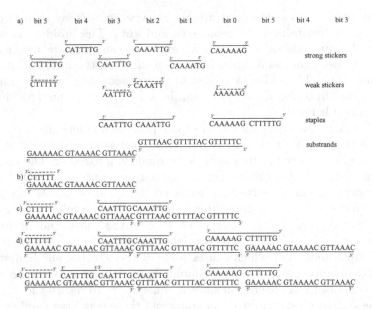

Fig. 1. Examples of Proposed Extended Sticker System

can be determined by counting Ts. The middle-bit positions (4 represented by GTAAAAC and 1 represented by GTTTTAC) of each substrand type may be used to store data; the other positions are intended to provide structural linkage. We will refer to the two substrand types in Fig. 1 as β and σ via two user-accessible system tubes, T_β and T_σ, respectively. (The additional α-type used by the $n_s = 3$ algorithm in the next section is not used here.) In addition, here $n_s = 2$ types of staples (each consisting of $2m$ bases) are provided: β-to-σ (also called BS which connects bit _BS=3 on β to bit BS_=2 on σ) and σ-to-β (also called SB which connects bits _SB=0 on σ to SB_=5 on β). The next section example uses different staple types and bit numbers.

Fig. 1 b) gives an example using a weak sticker on a β-type substrand (in tube T_L) as a result of setweak($T_L, 5$). As in the next section, the bit position could be symbolically referred to as SB_, (the side of the σ-to-β sticker that attaches to β), but as Fig. 1 is a low-level example, the 5 more clearly identifies the concrete operation on left-most bit of the left substrand (GAAAAAC). The utility of such a weak staple will be explained in the next section.

Fig. 1 c) continues the example using a staple to join two substrands together as a result of staple($T_L, 3, 2, T_\sigma$); the concatenated virtual strand result will be in tube T_L. The user must return unstapled σ substrands to T_σ; these can be identified by bit 2=0 (has neither staple nor sticker on GTTTAAC).

Fig. 1 d) gives a further example of staple($T_L, 0, 5, T_\beta$) which uses an additional β-type sticker to increase the length of the virtual strand. This causes a problem: the result has redundant GTAAAAC regions.

Fig. 1 e) shows how the novel solution proposed here is to "blot out" the GTAAAAC region in the virtual strand before the last staple operation, resulting in the structure shown. Because of the way sticker hardware operates, later operations will ignore the bit that was blotted out.

4   Log-Time Summation

This section will illustrate the extended-sticker system at a higher level than the low-level details of Fig. 1 using a simple application: adding a set of numbers. The conventional simple-sticker approach to add a pair of numbers [2] would be to have a single strand subdivided into three n-bit wide data fields, a (bits $(2n - 1)..n$), b (bits $(n - 1)..0$) and s=a+b (bits $(3n - 1)..2n$). To add more numbers would require an ever longer single strand. Instead, as illustrated in Fig. 2, bits $(3n - 1)..2n$ will be assigned to s substrands, bits $(2n - 1)..n$ will be assigned to a substrands and bits $(n - 1)..0$ will be assigned to b substrands. Prior to joining these together, they will reside in separate tubes.

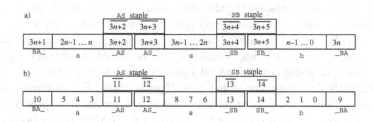

Fig. 2. Bit Layout in (asb) Virtual Strand. a) in general. b) $n = 3$.

Unlike folding single-strand origami [18], we ensure two substrands are involved by designing substrands to have *distinct* bits, numbered $3n$ or above, which will be assigned uniquely to receive staples for each substrand type. In the example, the simple linear data structure to be created has an a substrand stapled to an s substrand, which in turn is stapled to a b substrand, which might itself be stapled to another a substrand, etc. This means there are two extra bits in each substrand: one for stapling onto the left structure and another for stapling onto the right structure. The numbering of these bits that receive staples is arbitrary, for example, we could define _BA=$3n$, BA_=$3n + 1$, _AS=$3n + 2$, AS_=$3n + 3$, _SB=$3n + 4$ and SB_=$3n + 5$, where the underscore represents the substrand on which the specific bit number is located, and the letters represent the kind of staple which attaches to that bit position. For example, _BA is part of the b substrand whilst BA_ is the associated bit that is part of the a substrand. (Using contiguous numbers for user-level data bits helps simplify algorithmic design. The numbering is an arbitrary notation and does not necessarily describe the physical placement on the strand, as was done for simplicity in Fig. 1. Biochemical and physical considerations, like how the 3' and 5' sides of the substrands and staples come together, must be taken into account, but will be ignored here.) The operation `staple(t1,_BA,BA_,t0)` introduces to tube $t1$ staples (for joining a and b together), which consist of the Watson-Crick complements that represent bit addresses $3n$ and $3n + 1$. In addition to this BA-kind of staple, the addition example uses two other types: AS and SB. The staple-target bits for this example are: $3n$ and $3n + 5$ in b; $3n + 1$ and $3n + 2$ in a; and $3n + 3$ and $3n + 4$ in s. Fig. 2 shows the layout of bits in a simple example of a virtual strand.

Because the staple targets are ordinary bits, whether specific substrands accept staples is under programmed control. If a prior `set` (or `setweak`) occured on a particular substrand, the staple will be rejected (analogously to how a probe would be rejected). For a linear structure like this example, the number of staple types and the number of substrand types will be the same (n_s), and the number of bit positions dedicated to staples will be $2n_s$. For more complicated branching-data structures (such as might occur in DNA-encoding of large databases [4]), more staple types may be needed. A three-operand problem (like this addition example), needs at least three staple types.

In order to describe abstractly the values in each substrand, three distinct notations will be used. An uninitialized (all zero) substrand (from a vast supply in a system tube), will be denoted with the Greek equivalent to the name of the field; a field currently being processed will be denoted with the equivalent lower-case Roman italic; and a retired field (which has been blotted out) will be denoted with the equivalent upper-case Roman italic. For example, the field b may start uninitialized as β; then be shown as b whilst it is processed; and this same portion of the molecule will be shown as B after it has been blotted out. The extent of a virtual strand (constructed by zero or more staples) will be shown by a parenthesis. For example, (asb) shown in Fig. 2 is a virtual strand consisting of three substrands held together by two staples (AS and SB). This virtual strand can grow and have some of its substrands blotted out, for example, $(bASB)$ consists of four substrands held together by three staples (BA, AS and SB). Because of the blotted substrands, its value (as observable by `separate` operations) would be the same as the simple substrand (b).

The goal here is to compute the summation of all a_i and b_i substrands, i.e., $\sum_{i=0}^{N-1} a_i + \sum_{i=0}^{N-1} b_i$, assuming that each value is recorded in a single molecule. (A realistic system probably would record each value with redundant molecules and use refinement [7] to achieve Monte-Carlo distribution of errors [3]. Summing this is analogous to averaging the redundant values, with the effect of improving the accuracy of the Monte-Carlo arithmetic[16].) This paper ignores how the input a_i and b_i substrands were formed; for the large size of N considered here they could not have been individually input. Perhaps we could imagine [2] they are the result of previous sticker computation on random distributions shaped by a small number of input parameters. Many useful iterative algorithms [14] need much smaller input sets than the internal values they process.

Because of associativity and commutativity, there is an astronomical number of ways to compute partial sums yet yield the correct final sum. Of course, the most obvious linear-time approach could be used, in which one a value and one b value would be stapled onto a partial result at each stage. Given the slowness of biochemical operations, for N approaching the Avogadro constant, a linear-time approach would be much too slow to be feasible. Instead, consider a log-time approach, which uses the massive parallelism possible in the extended-sticker method. It will be convenient to start with the left operands (a substrands) in tube TL and the right operands (b substrands) in tube TR. To simplify the initial presentation of the algorithm, assume the number (N) of a substrands exactly matches the number of b substrands, and is a power of two. Of course, this is unrealistic as some unmatched, stray substrands will undoubtedly exist, but it is more convenient to ignore this at first. In addition to the input tubes TR and TL, system tubes TALPH, TBET and TSIG have a large supply of uninitialized substrands, and temporary tubes (like TLT and TRT) are available. The algorithm consists of $\lceil \log_2(N) \rceil$ iterations of Fig. 3.

```
            staple(TL,_AS,AS_,TSIG);
            separate3(TL,TSIG,TL,AS_);
            staple(TL,_SB,SB_,TR);
            fixstrays();
            for (i=0;i<n;i++) //"s=a+b"in TL..TL+3
                fulladd((2*n+i),i,(n+i),TL);
            combine(TL,(TL+2));
            for (i=0;i<2*n;i++)//blot "a,b"
                set(TL,i);
            split(TR,TL);
setweak(TL,BA_);                    setweak(TR,_BA);
staple(TL,_BA,BA_,TALPH);          staple(TR,BA_,_BA,TBET);
separate3(TL,TALPH,TL,BA_);        separate3(TR,TBET,TR,_BA);
clearweak(TL);/*BA_*/             clearweak(TR);/*_BA*/
for (i=0;i<n;i++)    //a=s in TL, b=s in TR, blot s in both
{ separate(TLT,TL,(i+2*n));        separate(TRT,TR,(i+2*n));
  set(TLT,(i+n));set(TL,(i+2*n));  set(TRT,i);set(TR,(i+2*n));
  combine(TL,TLT);                 combine(TR,TRT);}
```

Fig. 3. Log-time Summation Algorithm

4.1 Algorithm Trace

A summary of the first two iterations of the algorithm is given in Table 1. The algorithm starts by stapling (σ) substrands from a system tube to the (a) substrands in TL. Excess (σ) substrands are returned to the system tube. The (b) substrands in TR are combined with the ($a\sigma$) virtual strands in TL. These two substrands are stapled together, ignoring the fixstrays routine since we assume the number of (b) strands and ($a\sigma$) virtual strands match perfectly. The algorithm computes s=a+b using n iterations of the `fulladd` routine from [2]. (The `combine(TL,(TL+2))` is part of the algorithm from [2].) Any dyadic operation implementable with stickers could be used at this point; it is desirable for the operation to be commutative and at least approximately associative. (For example, real LNS addition [2], which needs a small number of scratch bits, could have these bits as part of an s substrand slightly-wider than a or b. Floating-point operations are also possible, but the number of scratch bits would be greater.) The input data, a and b, are blotted to A and B by setting bits $(2n-1)..0$, leaving only s with detectable data. At this point, the virtual strands (AsB) are split (for the sake of discussion, perfectly) in half between TL and TR.

From this point on, operations that are listed on the same line can be performed in parallel. If we left the target bits of the BA staple alone, an ambiguity would occur. For example, in tube TR, we want to staple (β) on the left of (AsB) using a BA staple. Unfortunately, both (β) and (AsB) have a _BA target bit. Without the `setweak`, sometimes ($AsBAsB$) would be formed instead of the desired (βAsB). A similar problem would occur in tube TL, where we want to staple (α) on the right of (AsB) using a BA staple. Using the `setweaks`, the algorithm temporarily forces right concatenation in TL and left concatenation in TR. Excess α and β substrands are returned to the system tubes, and the `clearweaks` will allow the structure to grow in the next iteration.

The reason for splitting into TL and TR is so that half of the s values can be copied to a substrands and the other half to b substrands. Once this is done, all s values can be blotted, producing ($ASBa$) in TL and ($bASB$) in TR, which are effectively equivalent to (a) and (b), respectively, which these tubes held at the beginning. Under the assumptions here, the number of ($ASBa$) and ($bASB$) will be exactly half the number of those initial inputs.

Table 1 shows iteration 2 of the same code, giving ($ASBASBASBa$) and ($bASBASBASB$). Both results are the sum of four numbers (the middle ASB segments held intermediate results).

4.2   Complexity

At the end of the jth iteration ($1 \leq j \leq \lceil\log_2(N)\rceil$), the combined number of virtual strands in TL and TR will be $2^{\lceil\log_2(N)\rceil-j}$, the number of inputs processed per virtual strand will be 2^j, and the number of substrands in each virtual strand will be $3\cdot(2^j-1)+1$ (of which only one, either an a or b, contains actual data). Considering parallism, the algorithm needs 9 time steps for statements outside loops. With n iterations of `fulladd` from [2] ($4n$ time steps using TL and three other tubes: TL+1 through TL+3, which could be shared with TR, TRT and TLT), $2n$ iterations of one statement to blot a and b ($2n$ time steps), and n iterations of the bottom loop to copy and blot s ($3n$ time steps), the algorithm uses four tubes and takes $\lceil\log_2(N)\rceil(9n+9)$ time steps. Suppose, hypothetically, we wish to add $N = 2^{79} \approx 1$ mole of 79-bit numbers contained in $n = 161$-bit substrands (large enough to hold any partial sum; unbounded multiprecision integers, as in Common Lisp, would require extra staple and substrand types). Assuming the time to staple remains the same as other operations as the virtual-strand lengths increase, we need $\lceil \log_2(N) \rceil (9n + 9) = 79 \cdot 1458 \approx 1.2 \cdot 10^5$ time steps. Using all of its parallel-processing abilities, the most powerful (60 petaFLOP) electronic supercomputer available in 2013 would require $6 \cdot 10^{26}/(6 \cdot 10^{16}) \approx 10^{10}$ seconds, or in other words, the biochemical supercomputer could outperform the electronic one as long as its time steps are faster than $10^{10}/(1.2 \cdot 10^5) \approx 8.3 \cdot 10^4$ seconds $\approx$ one day, which seems possible with margin to spare. Of course, adding a mole of numbers at this rate is only hypothetical, since 10^{10} seconds is over 300 years, and the DNA-based computation is unlikely to be reliable as described.

4.3   Macro, Micro and Hybrid Implementation

The preceding analysis assumes tubes large enough to hold all of the substrands at one time interconnected via macroscale tubes and pumps. A more promising alternative is microfluidics [8,6], which has been suggested for the simple sticker system [2,9] by partitioning the strands into (say 10^{-6} mol) droplets processed in parallel on several Micro-Electro-Mechanical System(MEMS) chips that also contain the electronic components to control the movement and processing of the droplets. Since the simple sticker system has no interstrand communication, such partitioning does not diminish its capabilities. The smaller size of the droplet suggests that reaction times[6], and therefore sticker time step will be reduced compared to a macroscale tube.

A completely microfluidic implementation of the extended sticker system will not be able to perform the log-time summation algorithm to completion. Instead, it would generate a partial sum on each MEMS chip. To overcome this, a hybrid electronic/biochemical approach could be used. For example, the proposed mole-scale computation could be carried out by one million MEMS chips (comparable to the number of chips in an electronic supercomputer) interconnected via a very simple electronic network. Assuming the electronic network could transmit one partial sum per microsecond, the problem could be completed in one second after each MEMS chip has converted its result from biochemical to electronic representation. The number of substrands processed on each individual MEMS chip is one millionth of the macroscale implementation. In general, let 2^{N_e} be the number of MEMS chips, $N_b = \lceil \log_2(N) \rceil - N_e$, t_e be the time to transmit one value electronically, t_c be the time to convert to electronic representation and t_b be the biochemical time step. The total time for the hybrid system is $t_e \cdot 2^{N_e} + t_c + t_b \cdot (9n + 9)N_b$. When t_e, t_c and t_b are constant, this does not vary by more than a factor of about two for $0 \leq N_e \leq N/2$, allowing the choice for N_e to based on MEMS cost and droplet size (with associated lower t_b). For the million-chip ($N_b = 59$) example, the first two terms can be ignored. Assuming the same biochemical clock of one day, the million-chip hybrid system would complete in a still unrealistic 235 years. The advantage, though, of the microfluidic implementation is that a much faster time step of a few minutes may be possible, perhaps allowing completion of the algorithm in under one year.

Table 1. Symbolic Trace

step	TL	TR
	(a)	(b)
`staple(TL,_AS,AS_,T`$_\sigma$`)`	$(a\sigma)$	(b)
`staple(TL,_SB,SB_,TR))`	$(a\sigma b)$	
`s=a+b`	(asb)	
`blot a,b`	(AsB)	
`split(TR,TL)`	(AsB)	(AsB)
`staple(TL,_BA,BA_,T`$_\alpha$`)`		
`staple(TR,_BA,BA_,T`$_\beta$`)`	$(AsB\alpha)$	(βAsB)
`a=s in TL;b=s in TR`	$(AsBa)$	$(bAsB)$
`blot s in TL,TR`	$(ASBa)$	$(bASB)$
`staple(TL,_AS,AS_,T`$_\sigma$`)`	$(ASBa\sigma)$	$(bASB)$
`staple(TL,_SB,SB_,TR)`	$(ASBa\sigma bASB)$	
`s=a+b`	$(ASBasbASB)$	
`blot a,b`	$(ASBAsBASB)$	
`split(TR,TL)`	$(ASBAsBASB)$	$(ASBAsBASB)$
`staple(TL,_BA,BA_,T`$_\alpha$`)`		
`staple(TR,_BA,BA_,T`$_\beta$`)`	$(ASBAsBASB\alpha)$	$(\beta ASBAsBASB)$
`a=s in TL;b=s in TR`	$(ASBAsBASBa)$	$(bASBAsBASB)$
`blot s in TL,TR`	$(ASBASBASBa)$	$(bASBASBASB)$

4.4 Stray Substrands

The discussion of the summation algorithm so far has ignored `fixstrays()`. This routine, shown in Fig. 4, is intended to deal with the likely situation that when the substrands in TL and TR are stapled together with SB staples, the number of substrands in the two tubes will not be exactly the same, and even if they were, some substrands may not staple together. To overcome this problem, `fixstrays()` uses TR (which at this point is either empty or has half-stapled substrands) and a temporary tube (TB) to take the stray strands which did not staple properly, and convert them into a format which will be compatible with later steps in the addition algorithm. The routine tests the SB_ bit of the b substrand; any substrands where this bit is zero did not receive a staple. Those substrands are transferred to TB. (In a similar way, any substrands where the _SB bit is zero are transferred to TR; it is likely strays would already have been transfered to TR as half-stapled exceptions from `staple(TL,_SB,SB_,RL)`.) At this point, steps involving TR and TB may be executed in parallel.

The process for TB is slightly more complicated. The first step involves taking the substrands that were just transferred into TB (which do not have a staple on their SB_ bit) and concatenating a new σ substrand. Because there are a surplus of these in the system tube, all the unstapled substrands from the previous step will now have σ substrand concatenated on the left. As in the earlier examples, the user must return the unused σ substrands to the system tube. Because the goal is to make the stickers of tube TB have the same format as if they were

```
fixstrays()
{                separate3(TL,TB,TL,SB_);
                 separate3(TL,TR,TL,_SB);
   staple(TB,SB_,_SB,TSIG);          staple(TR,_SB,SB_,TBET);
   separate3(TB,TSIG,TB,_SB);        separate3(TR,TBET,TR,SB_);
   staple(TB,AS_,_AS,TALPH);
   separate3(TB,TALPH,TB,_AS);
   combine(TL,TB);                   combine(TL,TR);  }
```

Fig. 4. Fixing Stray Substrands

the result of successful stapling in the calling routine, it is also necessary to concatenate an α substrand using an AS staple. Again, the unused α substrands need to be returned to the system tube.

The steps for tube TR are easier: concatenate a β substrand using an SB staple and return the unused β substrands to the system tube. Even though the majority of substrands already in TR are half-stapled exceptions, the user-level command staple(TR,_SB,SB_,TBET) insures they all become fully stapled (again because of the surplus of β substrands in the system tube). This also takes care of the few substrands that might not have had a staple. After the parallel processing of TR and TB is complete, the contents of both tubes are combined with TL, and the summation algorithm may resume.

Versions of the JavaScript and CUDA simple-sticker simulators [21] used in [2,3] (which represent each sticker bit as an actual bit in the computer) were augmented to simulate this algorithm with extended stickers. Since JavaScript is slow but has powerful string operations (similar to Java), the JavaScript extended-sticker simulator models each sticker bit as a lower-case character (or space if it is zero) and each staple bit as an upper-case character (limiting $k \leq 26$). This representation allows faithful modeling of half-staple exceptions, and the exponential growth of virtual strands.

In contrast, the high-speed CUDA code running on a GPU was modified to simulate only the non-blotted data as actual bits in the GPU's device memory. For efficiency, the blotted data is discarded after each iteration. This slightly less faithful modeling (the half-staple exception cannot occur, instead being caught by separate3(TL,TR,TL,_SB)) increases performance. Six distinct CUDA kernels are required to implement simple-sticker operations from the above algorithm on the GPU; the staple operation is modeled by linear-time scanning of arrays by the host CPU. A linear-time (Knuth/Fisher-Yates) shuffle [13] is used to generate random permutations of tubes T_L and T_R on each iteration, which exercises the fixstrays() routine extensively.

5   Conclusions

An extended-sticker system has been described in this paper that uses staples to hybridize to contiguous portions of two substrands, forming virtual strands. In contrast to the simple sticker system, which has all addressable bits in a single strand, different substrand types have distinct addressable bits. Using staples, the system can initiate concatenation of substrands. Using weak stickers to prevent undesired concatenation, the user may control whether concatenation occurs on the left or right and, if desired, whether the concatenated substrand is fresh (has bits that are zero). The weak stickers may then be melted away, to allow the virtual strand to grow further. As virtual strand length increases, redundant bits will occur. To overcome the ambiguity, this paper proposes blotting out old values (setting all bits in the old substrand prior to concatenating a similar-type substrand onto the virtual strand).

Using the proposed extended-sticker operations, a novel log-time summation algorithm is described. It works by splitting the virtual strands into two tubes. A virtual strand from one of these tubes is concatenated on the left of a new substrand; a virtual strand from the other tube is concatenated on the right of that new substrand. Once the three substrand types have been joined together, conventional sticker addition [2] produces a partial sum, and the operands used to produce this sum are blotted out. At the conclusion of one iteration, there are half as many virtual substrands. Although longer, the virtual substrands at the end of the iteration are in a format compatible with the input of the algorithm, which means the time complexity is logarithmic. In theory, for large data sets, such a log-time algorithm will outperform any electronic supercomputer conceivable in the near future. In reality, issues of reliability, which are a concern for the simple-sticker system [20], will be even more significant in the design of extended-sticker systems. For example, in the simple-sticker system, the probability of a misclassification is independent of the number of algorithm steps performed. In the extended sticker system, it is unclear whether this will remain so as the virtual strand length grows and many redundant bits have been blotted out. An additional concern is the exponential growth of virtual strand length, which may make handling the molecules more difficult than the simple-sticker system. Further research is needed to characterize the kinds of failure modes possible in the extended-sticker system, and their associated probabilities.

It is possible to imagine further extensions to the proposed system. For example, including weak staples would allow programmed control over whether a particular hybridization of substrands should be accepted. This would allow more complex algorithms than the simple summation described in this paper. It also could provide a means to control the exponential growth of virtual-strand length and a means to eliminate waste products.

Acknowledgements. The author appreciates the referees comments, especially about the need for the half-staple exception. Also, thanks to I. Martinez-Perez for his encouragement.

References

1. Adleman, L.: Molecular Computation of Solutions to Combinatorial Problems. Science 266, 1021–1024 (1994)
2. Arnold, M.G.: An Improved DNA-Sticker Addition Algorithm and Its Application to Logarithmic Arithmetic. In: Cardelli, L., Shih, W. (eds.) DNA 17. LNCS, vol. 6937, pp. 34–48. Springer, Heidelberg (2011)
3. Arnold, M.G.: Improved DNA-sticker Arithmetic: Tube-Encoded-Carry, Logarithmic Number System and Monte-Carlo Methods. Natural Computing (2012), doi: 10.1007/s11047-012-9356-3
4. Brijder, R., Gillis, J.J.M., Van den Bussche, J.: Graph-theoretic formalization of hybridization in DNA sticker complexes. In: Cardelli, L., Shih, W. (eds.) DNA 17. LNCS, vol. 6937, pp. 49–63. Springer, Heidelberg (2011)
5. Chang, W.-L., et al.: Fast Parallel Molecular Algorithms for DNA-Based Computation: Factoring Integers. IEEE Trans. Nanobiosci. 4, 149–163 (2005)
6. Chakrabarthy, K., et al.: Design Tools for Digital Microfluidic Biochips. IEEE Trans. Comp.-Aid. Des. 29, 1001–1017 (2010)
7. Chen, K., Winfree, E.: Error Correction in DNA Computing: Misclassification and Strand Loss. DNA-Based Computing 5, 49–63 (2000)
8. Fobel, R., Fobel, C., Wheeler, A.R.: Dropbot: An Open-Source Digital Microfluidic Control System with Precise Control of Electrostatic Driving Force and Instantaneous Drop Velocity Measurement. Appl. Phys. Lett. 102, 193513 (2013)
9. Grover, W.H.: Microfluidic Molecular Processors for Computation and Analysis. Ph.D. Dissertation, Chemistry Dept., University Of California, Berkeley (2006)
10. Guo, P., Zhang, H.: DNA Implementation of Arithmetic Operations. In: 2009 Int. Conf. Natural Comp., pp. 153–159 (2009)
11. Hartemink, A.J., Gifford, D.K.: Thermodynamic Simulation of Deoxyoligonucleotide Hybridization for DNA Computation. DIMACS 48, 15–25 (1999)
12. Ignatova, Z., Martinez-Perez, I., Zimmermann, K.: DNA Computing Models, Sec. 5.3. Springer, New York (2008)
13. Knuth, D.E.: The Art of Computer Programming: 2 Seminumerical Algorithms, pp. 124–125. Addison-Wesley, Reading (1969)
14. Makino, J., Taiji, M.: Scientific Simulations with Special-Purpose Computers—the GRAPE Systems. Wiley, Chichester (1998)
15. Martinez-Perez, I.M., Brandt, W., Wild, M., Zimmermann, K.: Bioinspired Parallel Algorithms for Maximum Clique Problems on FPGA Architectures. J. Sign Process. Syst. 58, 117–124 (2010)
16. Parker, S., Pierce, B., Eggert, P.R.: Monte Carlo Arithmetic: How to Gamble with Floating Point and Win. Computing in Science and Engineering 2(4), 58–68 (2000)
17. Qian, L., Winfree, E.: A Simple DNA Gate Motif for Synthesizing Large-scale Circuits. J. R. Soc. Interface (2011), doi:10.1098/rsif.2010.0729
18. Rothemund, P.K.W.: Folding DNA to Create Nanoscale Shapes and Patterns. Nature 440, 297–302 (2006)
19. Roweis, S., et al.: A Sticker-Based Model for DNA Computation. J. Comp. Bio. 5, 615–629 (1996)
20. Roweis, S., Winfree, E.: On Reduction of Errors in DNA Computation. J. Computational Bio. 6, 65–75 (1998)
21. JavaScript and CUDA DNA Sticker simulators, http://www.xlnsresearch.com/sticker.htm
22. Yang, X.Q., Liu, Z.: DNA Algorithm of Parallel Multiplication Based on Sticker Model. Comp. Engr. App. 43(16), 87–89 (2007)

Parallel Computation
Using Active Self-assembly*

Moya Chen, Doris Xin, and Damien Woods

California Institute of Technology
Pasadena, CA 91125, USA

Abstract. We study the computational complexity of the recently proposed nubots model of molecular-scale self-assembly. The model generalizes asynchronous cellular automaton to have non-local movement where large assemblies of molecules can be moved around, analogous to millions of molecular motors in animal muscle effecting the rapid movement of large arms and legs. We show that nubots is capable of simulating Boolean circuits of polylogarithmic depth and polynomial size, in only polylogarithmic expected time. In computational complexity terms, any problem from the complexity class NC is solved in polylogarithmic expected time on nubots that use a polynomial amount of workspace. Along the way, we give fast parallel algorithms for a number of problems including line growth, sorting, Boolean matrix multiplication and space-bounded Turing machine simulation, all using a constant number of nubot states (monomer types). Circuit depth is a well-studied notion of parallel time, and our result implies that nubots is a highly parallel model of computation in a formal sense. Thus, adding a movement primitive to an asynchronous non-deterministic cellular automaton, as in nubots, drastically increases its parallel processing abilities.

1 Introduction

We study the theory of molecular self-assembly, working within the recently-introduced *nubots* model by Woods, Chen, Goodfriend, Dabby, Winfree and Yin [43]. Do we really need another new model of self-assembly? Consider the biological process of embryonic development: a single cell growing into an organism of astounding complexity. Throughout this active, fast and robust process there is growth and movement. For example, at an early stage in the development of the fruit fly Drosophila, the embryo contains approximately 6,000 large cells arranged on its ellipsoid-shaped surface. Suddenly, within 4-minutes, the embryo changes shape to become invaginated, creating a large structure that becomes the mesoderm, and ultimately muscle. How does this fast rearrangement occur? A large fraction of these cells undergo a rapid, synchronized and highly parallel rearrangement of their internal structure where, in each cell, one end of

---

* Supported by National Science Foundation grants CCF-1219274, 0832824 (The Molecular Programming Project), and CCF-1162589. mpchen@caltech.edu, doris.s.xin@gmail.com, woods@caltech.edu. A full version of this paper will appear on the arXiv.

D. Soloveichik and B. Yurke (Eds.): DNA 2013, LNCS 8141, pp. 16–30, 2013.

the cell contracts and the other end expands. This is achieved by a mechanism that seems to crucially involve thousands of molecular-scale motors known as myosin pulling and pushing the cellular cytoskeleton to quickly effect this rearrangement [25]. At an abstract level one can imagine this as being analogous to how millions of molecular motors in a muscle, each taking a tiny step but acting in a highly parallel fashion, effect rapid long-distance muscle contraction. This rapid parallel movement, combined with the constraint of a fixed cellular volume, as well as variations in the elasticity properties of the cell membrane, can explain this key step in embryonic morphogenesis. Indeed, molecular motors that together, in parallel, produce macro-scale movement are a ubiquitous phenomenon in biology.

We wish to understand, at a high level of abstraction, the ultimate limitations and capabilities of such molecular scale rearrangement and growth. We do this by studying a theoretical model that includes these capabilities. As a first step towards such understanding, we show in this paper that large numbers of tiny motors (that can each pull or push a tiny amount) coupled with local state changes on a grid, are sufficient to quickly solve problems deemed to be inherently parallelizable. This result, described formally below in Section 1.2, demonstrates that our model, the nubots model, is a highly parallel computer in a computational complexity-theoretic sense.

Another motivation, and potential test-bed for our theoretical model and results, is the fabrication of active molecular-scale structures. Examples include DNA-based walkers, DNA origami that reconfigure, and simple structures called molecular motors [45] that transition between a small number of discrete states [43]. In these systems the interplay between structure and dynamics leads to behaviors and capabilities that are not seen in static structures, nor in other unstructured but active, well-mixed chemical reaction network type systems. Our theoretical results here, and those in [43], provide a sound basis to motivate the experimental investigation of large-scale active DNA nanostructures.

There are a number of theoretical models of molecular-scale algorithmic self-assembly processes [33]. For example, the abstract Tile Assembly Model, where individual square DNA tiles attach to a growing assembly lattice one at a time [41, 36, 17], or the two-handed (hierarchical) model where large multi-tile assemblies come together [1, 8, 12, 15], or the signal tile model where DNA origami tiles that form an "active" lattice with DNA strand displacement signals running along them [20, 30, 31], as well as models where one can program tile geometry [13, 18], temperature [1, 22, 39], concentration [6, 9, 16, 23] mixing stages [12, 14] and connectivity/flexibility [21].

The well-studied abstract Tile Assembly Model [41] is an asynchronous, and nondeterministic, cellular automaton with the restriction that state changes are irreversible and happen only along a crystal-like growth frontier. The nubots model is a generalization of an asynchronous and nondeterministic cellular automaton, where we have a non-local movement primitive. Nubots is intended to be a model of molecular-scale phenomena so it ignores friction and gravity, allows for the creation/destruction of monomers (we assume an invisible "fuel" source) and has a notion of Brownian motion (called agitation, but not used in this paper). Instances of the model evolve as continuous time Markov processes, hence time is modeled as in stochastic chemical kinetics [5, 38]. The style of movement in nubots is analogous to that seen in reconfigurable robotics [7, 37, 26], and indeed results in these robotics models show that non-local movement can be used to effect fast global reconfiguration [4, 3, 35]. The nubots model includes features seen in cellular automata, Lindenmayer systems [34] and graph grammars [24]. See [43] for more detailed comparisons with these models.

1.1   Previous Work on Active Self-assembly with Movement

Previous work on the nubots model [43] showed that it is capable of building large shapes and patterns exponentially quickly: e.g. lines and squares in time logarithmic in their size. Reference [43] goes on to describe a general scheme to build arbitrary computable (connected, 2D) size-n shapes in time and number of monomer states (types) that are polylogarthmic in n, plus the time and states required for Turing machine simulation due to the inherent algorithmic complexity of the shape. Furthermore, 2D patterns with at most n colored pixels, where the color of each pixel is computable in time $\log^{O(1)} n$ (i.e. polynomial in the length of the binary description of pixel indices), are nubots-computable in time and number of monomer types polylogarthmic in n [43]. The latter result is achieved without going outside the pattern boundary and in a completely asynchronous fashion. Many other models of self-assembly are not capable of this kind of parallelism. The goal of the present paper is to formalize the kind of parallelism seen in nubots via computational complexity of classical decision problems.

Dabby and Chen [11] study an insertion-based model, where monomers insert between, and push apart, other monomers. In this nice simplification of nubots they build length-n lines in $O(\log^3 n)$ expected time and $O(\log^2 n)$ monomer types in 1D. They also show relationships with regular and context-free languages, and give a design for implementation with DNA.

1.2   Our Results

In the nubots model a program is specified as a set of nubots rules $\mathcal{N}$ and is said to decide a language $L \subseteq \{0,1\}^*$ if, beginning with a word $x \in \{0,1\}^*$ encoded as a sequence of $|x|$ "binary monomers" at the origin, the system eventually reaches a configuration with the 1 monomer at the origin if $x \in L$, and 0 otherwise. Let NC denote the (well-known) class of problems solved by uniform polylogarthmic depth and polynomial size Boolean circuits.[1] Our main result is stated as follows.

Theorem 1. *For each language $L \in$ NC, there is a set of nubots rules $\mathcal{N}_L$ that decides L in polylogarthmic expected time, constant number of monomer states, and polynomial space in the input string length. Moreover, for $i \geq 1$, NC^i is contained in the class of languages decided by nubots running in $O(\log^{i+3} n)$ expected time, $O(1)$ monomer states, and polynomial space in input length n.*

---

[1] NC, or Nick's class, is named after Nicholas Pippenger.

NC problems are solved by circuits of shallow depth, hence they can be thought of as those problems that can be solved on a highly parallel architecture (simply run each layer of the circuit on a bunch of parallel processors, after polylog parallel steps we are done). NC is contained in P—problems solved by polynomial time Turing machines (this follows from the fact that NC circuits are of polynomial size). Problems in NC (or the analogous function class) include sorting, Boolean matrix multiplication, various kinds of maze solving and graph reachability, and integer addition, multiplication and division. Besides its circuit depth definition, NC has been characterized by a large number of other parallel models of computation including parallel random access machines, vector machines, and optical computers [19, 44, 42]. It is widely conjectured, but unproven, that NC is strictly contained in P. In particular, problems complete for P (such as Turing machine and cellular automata [29] prediction, context-free grammar membership and many others [19]) are believed to be "inherently sequential"—it is conjectured that these problems are not solvable by parallel computers that run for polylogarithmic time on a polynomial number of processors [19, 10].

Thus our main result gives a formal sense in which the nubots model is highly parallel: our proof gives a nubots algorithm to efficiently solve any highly parallelizable (NC) problem in polylogarithmic expected time and constant states. This stands in contrast to sequential machines like Turing machines, that cannot read all of an n-bit input string in polylogarithmic time, and "somewhat parallel" models like cellular automata and the abstract Tile Assembly Model, which can not have all of n bits influence a single bit decision in polylogarithmic time.

In order to obtain this result we give a number of novel nubots constructions. We show how to simulate function-computing logarithmic space deterministic Turing machines in only polylogarithmic expected time on nubots. We also show how to sort numbers in polylogarithmic expected time. Our sorting routine is used throughout our construction and is inspired by mechanisms such as gel electrophoresis that sort based on physical quantities (e.g. mass) [27]. We give a polylogarithmic expected time Boolean matrix multiplication algorithm, as well as a new line growing routine and a new synchronization (fast message passing) routine. All of these constructions are carried out using only a constant number of nubot monomers states and rules.

Previous results [43] on nubots were of the form: for each $n \in \mathbb{N}$ there is a set of nubot rules $\mathcal{N}_n$ (i.e. the number of rules is a function of n) to carry out some task parametrized by n (examples: quickly grow a line of length n, or an $n \times n$ square, grow some complicated computable pattern or shape whose size is parametrized by n, etc.). For each NC problems our main result here gives a *single* set of rules (i.e. of constant size), that works for all problem instances.

1.3   Future Work and Open Questions

The line growth algorithm in [43] runs in expected time $O(\log n)$, uses $O(\log n)$ states and space $O(n) \times O(1)$ from a single seed monomer. In our construction (see full paper) we give another line growth algorithm that runs in expected time $O(\log^2 n)$, uses $O(1)$ states and space $O(n) \times O(1)$ from a size $O(\log n)$ seed.

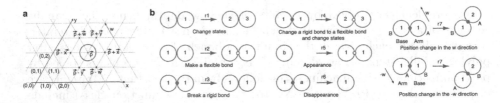

Fig. 1. Overview of nubots model. (a) A nubot configuration showing a single nubot monomer on the triangular grid. (b) Examples of nubot monomer rules. Rules r1-r6 are local cellular automaton-like rules, whereas r7 effects a non-local movement. A flexible bond is depicted as an empty red circle and a rigid bond is depicted as a solid red disk.

Is it possible to find a line-growth algorithm that does better than time × space × states = $\Omega(n \log^2 n)$?

Theorem 1 gives a lower bound on nubots power. What is the upper bound on confluent[2] polylogarthmic expected time nubots? One challenge involves finding better Turing machine space, or circuit depth, bounds on computing the movable set (see Section 2), and iterating this for many moves on a polynomial size (or larger) nubots grid.

Can we tighten our NC lower bound? Is the case that NC^k is contained in, say, the class of problems solved in $O(\log^{k+1} n)$ expected time on nubots? Our constructions make a lot of use of "synchronization" (where many monomers are simultaneously signaled to transition to a single common state), one way to improve our lower bound would be to see if it is possible to simulate circuits efficiently without using synchronization. The proof of Theorem 7.1 in [43] contains an example construction of a wide class of patterns that can be grown without synchronization. What conditions are necessary and sufficient for composition of arbitrary (unsynchronized) systems?

Is it possible to grow a structure of size $\Omega(n)$, in expected time $o(n)$ but without using the movement rule? Here the only source of movement comes from the "agitation" rule, which models the fact that in a liquid molecules are bombarding each other and jiggling all around. Every self-assembed molecular-scale structure was made under such conditions! Our question asks if we can *programmably exploit* this random molecular motion to build structures quicker than without it. Other open problems and further directions can be found in [43].

2   The Nubots Model and Other Definitions

In this section we formally define the nubots model. Figure 1 gives an overview of the model and rules, and Figure 2 gives an example of the movement rule.

The model uses a two-dimensional triangular grid with a coordinate system using axes x and y as shown in Figure 1(a). In the vector space induced by

---

[2] By confluent we mean a kind of determinism where the system (rules with the input) is assumed to always make a unique single terminal assembly.

this coordinate system, the *axial directions* $\mathcal{D} = \{\pm\overrightarrow{w}, \pm\overrightarrow{x}, \pm\overrightarrow{y}\}$ are the unit vectors along the grid axes. A pair $\overrightarrow{p} \in \mathbb{Z}^2$ is called a *grid point* and has the set of six *neighbors* $\{\overrightarrow{p} + \overrightarrow{u} \mid \overrightarrow{u} \in \mathcal{D}\}$. Let S be a finite set of monomer states. A nubot *monomer* is a pair $X = (s_i, p(X))$ where $s_i \in S$ is a state and $p(X) \in \mathbb{Z}^2$ is a grid point. Two monomers on neighboring grid points are either connected by a *flexible* or *rigid* bond, or else have no bond (called a *null* bond). Bonds are described in more detail below. A *configuration* C is a finite set of monomers along with the bonds between them.

One configuration *transitions* to another via the application of a single *rule*, $r = (s1, s2, b, \overrightarrow{u}) \rightarrow (s1', s2', b', \overrightarrow{u}')$ that acts on one or two monomers.[3] The left and right sides of the arrow respectively represent the contents of the two monomer positions before and after the application of rule r. Here $s1, s2 \in S \cup \{\text{empty}\}$ are monomer states where at most one of $s1, s2$ is empty (denotes lack of a monomer), $b \in \{\text{flexible}, \text{rigid}, \text{null}\}$ is the bond type between them, and $\overrightarrow{u} \in \mathcal{D}$ is the relative position of the $s2$ monomer to the $s1$ monomer. If either of $s1$ or $s2$ (respectively $s1'$ or $s2'$) is empty then b (respectively b') is null. The right is defined similarly, although there are some further restrictions on valid rules (involving $\overrightarrow{u}'$) described below. A rule is only applicable in the orientation specified by $\overrightarrow{u}$, and so rules are not rotationally invariant.

A rule may involve a movement (translation), or not. First, in the case of no movement: $\overrightarrow{u} = \overrightarrow{u}'$. Thus we have a rule of the form $r = (s1, s2, b, \overrightarrow{u}) \rightarrow (s1', s2', b', \overrightarrow{u})$. From above, at most one of $s1, s2$ is empty, hence we disallow spontaneous generation of monomers from empty space. *State change* and *bond change* occurs in a straightforward way, examples are shown in Figure 1(b). If $s_i \in \{s1, s2\}$ is empty and s_i' is not, then the rule induces the *appearance* of a new monomer at the empty location specified by $\overrightarrow{u}$ if $s2 = \text{empty}$, or $-\overrightarrow{u}$ if $s1 = \text{empty}$. If one or both monomers go from non-empty to empty, the rule induces the *disappearance* of monomer(s) at the orientation(s) given by $\overrightarrow{u}$.

For a *movement* rule it must be the case that $\overrightarrow{u} \neq \overrightarrow{u}'$ and $d(\overrightarrow{u}, \overrightarrow{u}') = 1$, where $d(u, v)$ is Manhattan distance on the triangular grid, and $s1, s2, s1', s2' \in S \setminus \{\text{empty}\}$. If we fix $\overrightarrow{u} \in \mathcal{D}$, then there are two $\overrightarrow{u}' \in \mathcal{D}$ that satisfy $d(\overrightarrow{u}, \overrightarrow{u}') = 1$. A movement rule is applied both (i) locally and (ii) globally, as follows.

(i) Locally, one of the two monomers is chosen nondeterministically to be the *base* (which remains stationary), the other is the *arm* (which moves). If the $s2$ monomer, denoted X, is chosen as the arm then X moves from its current position $p(X)$ to a new position $p(X) - \overrightarrow{u} + \overrightarrow{u}'$. After this movement $\overrightarrow{u}'$ is the relative position of the $s2'$ monomer to the $s1'$ monomer, as illustrated in Figure 1(b). Analogously, if the $s1$ monomer, Y, is chosen as the arm then Y moves from $p(Y)$ to $p(Y) + \overrightarrow{u} - \overrightarrow{u}'$. Again, $\overrightarrow{u}'$ is the relative position of the $s2'$ monomer to the $s1'$ monomer. Bonds and states may change during the movement.

---

[3] In reference [43] the nubots model includes "agitation": each monomer is repeatedly subjected to random movements that are intended to model Brownian motion and other uncontrolled fluid flows and movement. Our constructions work with or without agitation, hence they are robust to random uncontrolled movements, but we choose to ignore this issue and not formally define agitation for ease of presentation.

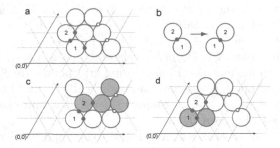

Fig. 2. An example of a movement rule with two results depending on the choice of arm or base. (a) Initial configuration. (b) Movement rule. (c) Result if the monomer with state 1 is the base. (d) Result if the monomer with state 2 is the base. We can think of (c) as pushing and (d) as pulling. Also, the affect on a flexible bonds (hollow red circles) and null bonds are shown.

(ii) Globally, the movement rule may push, or pull other monomers, or if it can do neither then it is not applicable. This is formalized as follows, and an example is shown in Figure 2. Let $\overrightarrow{v} \in \mathcal{D}$ be a unit vector. The $\overrightarrow{v}$-boundary of a set of monomers S is defined to be the set of grid points outside S that are unit distance in the $\overrightarrow{v}$ direction from monomers in S. Let C be a configuration containing adjacent monomers A and B, and let C' be C except that the bond between A and B is null in C' if not null in C. The *movable set* $M = \mathcal{M}(C, A, B, \overrightarrow{v})$ is the smallest subset of C' that contains A but not B and can be translated by $\overrightarrow{v}$ to give the set $M_{+\overrightarrow{v}}$ where the new configuration $C'' = (C' \setminus M) \cup M_{+\overrightarrow{v}}$ is such that: (a) monomer pairs in C' that are joined by rigid bonds have the same relative position in C'', (b) monomer pairs in C' that are joined by flexible bonds are neighbors in C'', and (c) the $\overrightarrow{v}$-boundary of M contains no monomers.

   If $\mathcal{M}(C, A, B, \overrightarrow{v}) \neq \{\}$, then the movement where A is the arm (which should be translated by $\overrightarrow{v}$) and B is the base (which should not be translated) is applied as follows: (1) the movable set $\mathcal{M}(C, A, B, \overrightarrow{v})$ moves unit distance along $\overrightarrow{v}$; (2) the states of, and the bond between, A and B are updated according to the rule; (3) the states of all the monomers besides A and B remain unchanged and pairwise bonds remain intact (although monomer positions and flexible/null bond orientations may change). If $\mathcal{M}(C, A, B, \overrightarrow{v}) = \{\}$, the movement rule is inapplicable (the rule is "blocked" and thus A is prevented from translating).

   An *assembly system* $T = (C_0, \mathcal{N})$ is a pair where C_0 is the initial configuration, and $\mathcal{N}$ is the set of rules. If configuration C_i transitions to C_j by some rule $r \in \mathcal{N}$, we write $C_i \vdash_\mathcal{N} C_j$. A *trajectory* is a finite sequence of configurations $C_1, C_2, \ldots, C_k$ where $C_i \vdash_\mathcal{N} C_{i+1}$ and $1 \leq i \leq k - 1$. An assembly system evolves as a continuous time Markov process. The rate for each rule application is 1. If there are k applicable transitions for C_i then the probability of any given transition being applied is $1/k$, and the time until the next transition is applied is an exponential random variable with rate k (i.e. the expected time is

$1/k)$.[4] The probability of a trajectory is then the product of the probabilities of each of the transitions along the trajectory, and the expected time of a trajectory is the sum of the expected times of each transition in the trajectory. Thus, $\sum_{t \in \mathcal{T}} \Pr[t] \text{time}(t)$ is the expected time for the system to evolve from configuration C_i to configuration C_j, where $\mathcal{T}$ is the set of all trajectories from C_i to any configuration isomorphic to C_j, that do not pass through any other configuration isomorphic to C_j, and $\text{time}(t)$ is the expected time for trajectory t.

2.1 Nubots and Decision Problems

Let $\mathbb{N} = \{0, 1, 2, \ldots\}$. Given a binary string $x \in \{0, 1\}^*$, written $x = x_0 x_1 \ldots x_{k-1}$, we let $\widetilde{x}$ denote a horizontal line of k nubot monomers that represent x using one of two "binary" monomer states. We let $|\widetilde{x}| \in \mathbb{N}$ denote the number of monomers in $\widetilde{x}$. Given a line of monomers A composed of m (previously defined) line segments, the notation $[A, i]$ means segment i of A, and $[A, i]_j$ means bit j of segment i of A. We next define what it means to decide a language (or problem) with nubots.

Definition 1. *A finite set of nubot rules $\mathcal{N}_L$ decides a language $L \subseteq \{0, 1\}^*$ if for all $x \in \{0, 1\}^*$ there is an initial configuration C_0 consisting of exactly the line $\widetilde{x}$ of monomers, positioned so that the left extent of $\widetilde{x}$ is at the origin $(0, 0)$, where by applying the rule set $\mathcal{N}_L$, the system always eventually reaches a configuration where there is an "answer" monomer at the origin in one of two states: (a) "accept" if $x \in L$, or (b) "reject" if $x \notin L$. Further, from the time it first appears, the answer monomer never changes state.*

2.2 Boolean Circuits and the Class NC

We define a Boolean circuit to be a directed acyclic graph, where the nodes are called gates and each node has a label that is one of: input (with in-degree 0), constant 0 (in-degree 0), constant 1 (in-degree 0), $\vee$ (OR, in-degree 1 or 2), $\wedge$ (AND, in-degree 1 or 2), $\neg$ (NOT, in-degree 1). One of the gates is also identified as the output gate. The *depth* of a circuit is the length of the longest path from an input gate to the output gate. The *size* of a circuit is the number of gates it contains. A circuit computes a Boolean (yes/no) function on a fixed number of Boolean variables, by the inputs and constants defining the output gate value in the standard way. In order to compute functions over an arbitrarily number of variables, we define (usually, infinite) families of circuits. We say that a family of circuits $\mathcal{C}_L = \{c_n \mid c_n \text{ is a circuit with } n \in \mathbb{N} \text{ input gates}\}$ decides a language $L \subseteq \{0, 1\}^*$ if for each $x \in \{0, 1\}^*$ circuit $c_{|x|} \in \mathcal{C}_L$ on input x outputs 1 if $w \in L$ and 0 if $w \notin L$.

In a *non-uniform* family of circuits there is no required similarity, or relationship, between family members. We use a *uniformity function* that algorithmically specifies some similarity between members of a circuit family. Roughly speaking,

---

[4] For simplicity, when counting the number of applicable rules for a configuration, a movement rule is counted twice, to account for the two choices of arm and base.

a *uniform circuit family* C is an infinite sequence of circuits with an associated function $f : \{1\}^* \to C$ that generates members of the family and is computable within some resource bound. Here we care about logspace-uniform circuit families:

Definition 2 (L-uniform circuit family). *A circuit family C is L-uniform, if there is function $f : \{1\}^* \to C$ that is computable on a deterministic logarithmic space Turing machine, and where $f(1^n) = c_n$ for all $n \in \mathbb{N}$, and $c_n \in C$ is a description of a circuit with n input gates.*

Without going into details, we assume reasonable encodings of circuits as strings. There are stricter, but more technical to state, notions of uniformity used in the literature [2, 46, 19, 28] (which we do not require since we are giving a lower bound on power), and circuit classes are reasonably robust under these more restrictive definitions.

Define NC^i to be the class of all languages $L \subseteq \{0,1\}^*$ that are decided by $O(\log^i n)$ depth, polynomial size L-uniform Boolean circuit families. Define $\mathrm{NC} = \bigcup_{i=0}^{\infty} \mathrm{NC}^i$, in other words NC is the class of languages decided by poly-logarithmic depth and polynomial size L-uniform Boolean circuit families. Since NC circuits are of polynomial size, they can be simulated by polynomial time Turing machines, and so $\mathrm{NC} \subseteq \mathrm{P}$. It remains open whether this containment is strict [19]. See [40] for more on circuits.

3 Proof Overview of Theorem 1

Here we give a high-level overview of the proof of Theorem 1. The full paper contains the detailed proof, which includes novel parallel nubots algorithms for line growth, sorting, Boolean matrix multiplication, space bounded function-computing Turing machine simulation, parallel function evaluation for functions of a certain form, Boolean circuit generation, and Boolean circuit simulation.

For each language $L \in \mathrm{NC}$, we show that there exists a finite set of nubots rules $\mathcal{N}_L$ that decides L in the sense of Definition 1. Let $\mathcal{C}_L$ be the circuit family that decides L. We begin with the observation that since L is in logspace-uniform NC, there is a deterministic Turing machine $\mathcal{M}_L$ that uses logarithmic space (in its input size) such that on unary input 1^n, $\mathcal{M}_L(1^n) = c_n$, where c_n is a description of the unique circuit in $\mathcal{C}_L$ that has n input gates.

Our initial nubots configuration consists of a length-n line of binary nubots monomers denoted $\widetilde{x}$ that represents some $x \in \{0,1\}^*$, and is located at the origin. From this we create another length-n line of monomers that encode the unary string 1^n to be used for the creation of the circuit c_n. The rule set $\mathcal{N}_L$ includes a description of $\mathcal{M}_L$. At a very abstract level, the system first generates a circuit by simulating the computation of $\mathcal{M}_L$ on input 1^n, and producing a nubots configuration (collection of monomers in a connected component) that represents the circuit c_n. The circuit is then simulated on input x. Both of these tasks present a number of challenges.

3.1 Circuit Generation

Here we describe the fast parallel simulation of the logspace machine $\mathcal{M}_L$. Logspace Turing machines have a read-only input tape with n input symbols, a read-write worktape of length $O(\log n)$, and a write-only output tape where the output tape head is assumed to always move right after writing a symbol. A configuration consists of the input tape head position, worktape contents, and worktape head position. There are at most $O(n^c)$ distinct configurations of this form, for some $c \in \mathbb{N}$, which comes from the $O(\log n)$ bound on the worktape length. Hence $\mathcal{M}_L$ runs in time $O(n^c)$. We assume that $\mathcal{M}_L$ stops in a *halt* state (we are simulating a halting, function-computing, deterministic machine, so it can be assumed to always halt in a special *halt* state). As noted, $\mathcal{M}_L$ runs in time $O(n^c)$, however we require a nubots simulation that runs in expected time that is merely polylogarithmic in n. To achieve this our simulation of $\mathcal{M}_L$ works in a highly parallel fashion, described below.

First, we describe the adjacency matrix of the configuration graph G of $\mathcal{M}_L$ on input 1^n. A configuration graph G is a directed graph, where each node represents a configuration of $\mathcal{M}_L$ on (the fixed) input 1^n [32]. There is an edge from node i to node j if and only if $\mathcal{M}_L$ moves from configuration i to configuration j in a single step. From the previously-mentioned basic facts about logspace machines, the number of nodes in G is at most polynomial in n. Further, nodes in G have out degree 0 or 1 ($\mathcal{M}_L$ is deterministic), the "halt" node has out degree 0 (we assume there are no transitions out of the halt state), and there a unique halt configuration ($\mathcal{M}_L$ completes its computation by wiping the worktape, returning all tape heads to the beginning of their tapes, and entering the halt state). The nubots system $\mathcal{N}_L$ begins by generating a representation of the adjacency matrix of graph G of machine $\mathcal{M}_L$ on input 1^n. This is achieved by building a "counter," that grows from the n monomers (that encode 1^n) to become an $O(n^c) \times O(\log n)$ rectangle, the rows of which enumerate the syntactically correct configurations of the machine via the known time ($O(n^c)$) and space ($O(\log n)$) bounds (some of these configurations are reachable on this input, and some are not). The list of configurations are grown in expected time $O(\log^2 n)$, polynomial space and only $O(1)$ monomer states. We then make a copy of this list, and pairwise compare every entry in the copy to that of the original—a process achieved via iterative copying of the list along with some geometric rearrangement tricks. The comparisons are done in parallel, where for each i, j it is checked whether configuration j is reachable from configuration i in one step on $\mathcal{M}_L$ (each such comparison depends only on configurations i, j and so can be computed in expected time $O(\log n)$ since the nubot rules $\mathcal{N}_L$ directly encode the program $\mathcal{M}_L$). The result of this process is quickly (in parallel) rearranged to form a new list (a line of monomers) that encodes the result of all of these comparisons, and thus represents the entire binary adjacency matrix M_G.

After the adjacency matrix M_G is constructed, the nubots system computes reachability on the graph G. Specifically, the rules $\mathcal{N}_L$ compute whether a path exists from the node representing the initial configuration of $\mathcal{M}_L$ on input x to the node representing the unique halting configuration in the halt state.

However, this directed graph is of size polynomial in n, so a sequential algorithm would be too slow for our purposes. We quickly (in polylog expected time) solve this reachability problem by parallel iterated matrix squaring of the adjacency matrix M_G. More precisely, we iterate $M_G := M_G^2 + M_G$ a total of $O(\log n)$ times to give the matrix M_G'. The column in M_G' that represents the halt node of graph G contains non-zero entries for exactly those nodes that have a path to the halt node [32]. The Boolean matrix squaring is carried out as follows. M_G is represented as a line of monomers, this line iscopied, and every entry of the two copies is pairwise ANDed, this involves further copying and geometric arrangement. The results are rearranged (using a novel nubots sorting algorithm discussed below) and then ORed in parallel to give the Boolean matrix M_G^2.

This parallel matrix multiplication algorithm constitutes the main part of a construction to simulate a logspace Turing machine that *decides* some language (if we also take account of accept/reject states). However, here we wish to simulate a machine that computes a *function*: $\mathcal{M}_L$'s output is a description of the circuit c_n, so we are not yet done. We add the following assumption to $\mathcal{M}_L$: it has a counter on its worktape that starts at 0 and is immediately incremented each time $\mathcal{M}_L$ writes to the output tape (this counter merely adds a $O(\log n)$ term to the worktape length). Thus only the "output-producing" configurations involve a counter incrementation. We extract from the matrix M_G' exactly those configurations that satisfy the following two criteria: (1) they are on a path from the input configuration to the halt configuration (2) they produce output. To find (1) we simply filter out those nodes (configurations) that correspond to non-zero entries in both the row of the initial node, and the column of the halt node. To get (2) we sort, via a novel, fast parallel sorting algorithm (discussed below), these configurations in increasing order of the values on their workspace counters. Then we take this sorted list, and delete everything (in parallel) except the encoded output tape write symbol from each configuration. We use the counter to sort the write symbols and are left with a line of $\ell = O(n^k)$ monomers that represent the length-ℓ output tape of $\mathcal{M}_L$ on input 1^n. This line of monomers, which we denote $\widehat{c_n}$, is an encoding of the circuit c_n.

The line of monomers $\widehat{c_n}$ is next geometrically rearranged for fast parallel circuit simulation. Here, $\widehat{c_n}$ reorganizes itself into a ladder-like form as shown in Figure 3(c) via fast parallel folding. Each layer i of the circuit c_n as shown in Figures 3(a) is encoded as a row of nubot monomers, as shown in Figure 3(c) (our circuits are assumed to be layered [40]). The circuit is now ready to be simulated.

3.2   Circuit Simulation

Recall that the circuit input bits (encoded as binary monomers) are located at the origin, and that the entire circuit was "grown" from them. These monomers move to the first (bottom) row of the encoded circuit (Figure 3(c)) and position themselves so that each gate can "read" its 1 or 2 input bits. The jth gate on layer $i \geq 1$, is simulated by a single nubot monomer that reads its adjacent 1 or 2 input bits and then sends its "result bit" to the blue "wire address" regions directly above it (Figure 3(d), in blue). After each gate computes its bit, layer i "synchronizes" via a logarithmic in n expected time message passing

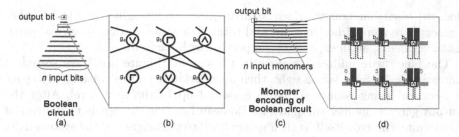

output bit

n input bits

Boolean
circuit

(a)

(b)

output bit

n input monomers

Monomer
encoding of
Boolean circuit

(c)

(d)

Fig. 3. Encoding of a Boolean circuit as a nubots configuration. (a) Boolean circuit with (b) detailed zoom-in. (c) Nubots configuration encoding the circuit, with zoom-in shown in (d). A wire leading out of a gate in (b) has a destination gate number encoded in (d) as strips of $O(\log n)$ blue binary monomers (indices in red). After a gate computes some Boolean function (one of $\vee$, $\wedge$, $\neg$) the resulting bit is tagged onto the relevant blue strip of monomers that encode the destination addresses (red numbers). Circuits are not necessarily planar, so to handle wire crossovers these result bits are first sorted in parallel based on their wire address, and then pushed up to the next layer of gates.

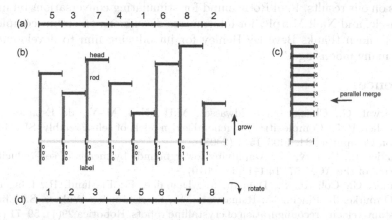

(a) 5 3 7 4 1 6 8 2

(b) head
     rod

     grow

     0 0 0   1 0 0   0 1 0   1 1 0   0 0 1   1 0 1   0 1 1   1 1 1
     label

(c) 8
     7
     6
     5
     4
     3 parallel merge
     2
     1

(d) 1 2 3 4 5 6 7 8   rotate

Fig. 4. High-level overview of the sorting algorithm. (a) A line of $m\lceil \log m \rceil$ monomers, split into m blue line segments ("heads") each is the binary representation of a natural number $i \le m$. (b) A blue head that encodes value i is grown to height i by a green rod. Purple "labels" are also grown at the bottom. (c) The heads are horizontally merged, using the labels to synchronize, to be vertically aligned. (d) Merged heads rotate down into a line configuration, giving the sorted list. Each stage occurs in expected time polylogarithmic in m. See full paper for details.

algorithm [43]. Next, we wish to send the "result" bits from layer i to layer $i+1$. Circuits are not necessarily planar, so we need to deal with wire crossings.

Wire crossings are handled via a fast parallel sorting routine (also used in earlier parts of the construction) that is loosely inspired by Murphy et al [27] who show that physical techniques, such as gel electrophoresis, can be used to sort numbers that are represented as the magnitude of some physical quantity. The sorting routine is illustrated in Figure 4. It takes as input a line of $m\lceil \log_2 m \rceil$ monomers, which is composed of m line segments each encoding a number in

$\lceil \log_2 m \rceil$ binary monomers. Each segment grows to a height equal to its value, segments are merged horizontally, and rotated down to vertical to give a sorted list of segments, all in expected time polylogarithmic in m.

The blue "wire address" regions in the circuit (Figure 3(d)) are sorted in increasing order from left to right, then appropriately padded with empty space in between (using counters), and are passed up to the next level. After the "output gate" monomer computes its Boolean function, it signals to the rest of the circuit to destroy itself. It then moves itself to the origin and the system halts (no more rules are applicable). This completes the overview of the simulation.

This overview ignores many details. In particular the nubots model is asynchronous, that is, rule updates happen independently via stochastic chemical kinetics. The construction includes a large number of synchronization steps and signal passing to ensure that all parts of the construction are appropriately staged, but yet the construction is free to carry out many fast, asynchronous, parallel steps between these "sequential" synchronization steps.

Acknowledgments. We thank Erik Winfree for valuable discussion and suggestions on our results, Paul Rothemund for stimulating conversations on molecular muscle, and Niall Murphy for informative discussions on circuit complexity theory. Damien thanks Beverley Henley for introducing him to developmental biology many moons ago.

References

1. Aggarwal, G., Cheng, Q., Goldwasser, M.H., Kao, M.-Y., de Espanes, P.M., Schweller, R.T.: Complexities for generalized models of self-assembly. SIAM Journal on Computing 34, 1493–1515 (2005)
2. Allender, E., Koucký, M.: Amplifying lower bounds by means of self-reducibility. Journal of the ACM 57, 14:1–14:36 (2010)
3. Aloupis, G., Collette, S., Damian, M., Demaine, E., Flatland, R., Langerman, S., O'rourke, J., Pinciu, V., Ramaswami, S., Sacristán, V., Wuhrer, S.: Efficient constant-velocity reconfiguration of crystalline robots. Robotica 29(1), 59–71 (2011)
4. Aloupis, G., Collette, S., Demaine, E.D., Langerman, S., Sacristán, V., Wuhrer, S.: Reconfiguration of cube-style modular robots using $O(\log n)$ parallel moves. In: Hong, S.-H., Nagamochi, H., Fukunaga, T. (eds.) ISAAC 2008. LNCS, vol. 5369, pp. 342–353. Springer, Heidelberg (2008)
5. Angluin, D., Aspnes, J., Eisenstat, D.: Fast computation by population protocols with a leader. In: Dolev, S. (ed.) DISC 2006. LNCS, vol. 4167, pp. 61–75. Springer, Heidelberg (2006)
6. Becker, F., Rapaport, I., Rémila, É.: Self-assemblying classes of shapes with a minimum number of tiles, and in optimal time. In: Arun-Kumar, S., Garg, N. (eds.) FSTTCS 2006. LNCS, vol. 4337, pp. 45–56. Springer, Heidelberg (2006)
7. Butler, Z., Fitch, R., Rus, D.: Distributed control for unit-compressible robots: goal-recognition, locomotion, and splitting. IEEE/ASME Transactions on Mechatronics 7, 418–430 (2002)
8. Cannon, S., Demaine, E.D., Demaine, M.L., Eisenstat, S., Patitz, M.J., Schweller, R.T., Summers, S.M., Winslow, A.: Two hands are better than one (up to constant factors): Self-assembly In the 2HAM vs. aTAM. In: STACS: 30th International Symposium on Theoretical Aspects of Computer Science, pp. 172–184 (2013)

9. Chandran, H., Gopalkrishnan, N., Reif, J.: Tile complexity of approximate squares. Algorithmica, 1–17 (2012)

10. Condon, A.: A theory of strict P-completeness. Computational Complexity 4(3), 220–241 (1994)

11. Dabby, N., Chen, H.-L.: Active self-assembly of simple units using an insertion primitive. In: SODA: Proceedings of the Twenty-fourth Annual ACM-SIAM Symposium on Discrete Algorithms, pp. 1526–1536 (January 2012)

12. Demaine, E., Demaine, M., Fekete, S., Ishaque, M., Rafalin, E., Schweller, R., Souvaine, D.: Staged self-assembly: nanomanufacture of arbitrary shapes with $O(1)$ glues. Natural Computing 7(3), 347–370 (2008)

13. Demaine, E.D., Demaine, M.L., Fekete, S.P., Patitz, M.J., Schweller, R.T., Winslow, A., Woods, D.: One tile to rule them all: Simulating any Turing machine, tile assembly system, or tiling system with a single puzzle piece (December 2012), Arxiv preprint arXiv:1212.4756 [cs.DS]

14. Demaine, E.D., Eisenstat, S., Ishaque, M., Winslow, A.: One-dimensional staged self-assembly. In: Cardelli, L., Shih, W. (eds.) DNA 17 2011. LNCS, vol. 6937, pp. 100–114. Springer, Heidelberg (2011)

15. Demaine, E.D., Patitz, M.J., Rogers, T.A., Schweller, R.T., Summers, S.M., Woods, D.: The two-handed tile assembly model is not intrinsically universal. In: Fomin, F.V., Freivalds, R., Kwiatkowska, M., Peleg, D. (eds.) ICALP 2013, Part I. LNCS, vol. 7965, pp. 400–412. Springer, Heidelberg (2013)

16. Doty, D.: Randomized self-assembly for exact shapes. SICOMP 39, 3521 (2010)

17. Doty, D., Lutz, J.H., Patitz, M.J., Schweller, R.T., Summers, S.M., Woods, D.: The tile assembly model is intrinsically universal. In: Proceedings of the 53rd Annual IEEE Symposium on Foundations of Computer Science, pp. 439–446 (October 2012)

18. Fu, B., Patitz, M.J., Schweller, R.T., Sheline, R.: Self-assembly with geometric tiles. In: Czumaj, A., Mehlhorn, K., Pitts, A., Wattenhofer, R. (eds.) ICALP 2012, Part I. LNCS, vol. 7391, pp. 714–725. Springer, Heidelberg (2012)

19. Greenlaw, R., Hoover, H.J., Ruzzo, W.L.: Limits to parallel computation: P-completeness theory. Oxford University Press, USA (1995)

20. Jonoska, N., Karpenko, D.: Active tile self-assembly, self-similar structures and recursion (2012), Arxiv preprint arXiv:1211.3085 [cs.ET]

21. Jonoska, N., McColm, G.L.: Complexity classes for self-assembling flexible tiles. Theoretical Computer Science 410(4), 332–346 (2009)

22. Kao, M., Schweller, R.: Reducing tile complexity for self-assembly through temperature programming. In: Proceedings of the Seventeenth Annual ACM-SIAM Symposium on Discrete Algorithm, pp. 571–580. ACM (2006)

23. Kao, M.-Y., Schweller, R.T.: Randomized self-assembly for approximate shapes. In: Aceto, L., Damgård, I., Goldberg, L.A., Halldórsson, M.M., Ingólfsdóttir, A., Walukiewicz, I. (eds.) ICALP 2008, Part I. LNCS, vol. 5125, pp. 370–384. Springer, Heidelberg (2008)

24. Klavins, E.: Directed self-assembly using graph grammars. In: Foundations of Nanoscience: Self Assembled Architectures and Devices, Snowbird, UT (2004)

25. Martin, A.C., Kaschube, M., Wieschaus, E.F.: Pulsed contractions of an actin–myosin network drive apical constriction. Nature 457(7228), 495–499 (2008)

26. Murata, S., Kurokawa, H.: Self-reconfigurable robots. IEEE Robotics & Automation Magazine 14(1), 71–78 (2007)

27. Murphy, N., Naughton, T.J., Woods, D., Henley, B., McDermott, K., Duffy, E., van der Burgt, P.J., Woods, N.: Implementations of a model of physical sorting. International Journal of Unconventional Computing 4(1), 3–12 (2008)

28. Murphy, N., Woods, D.: AND and/or OR: Uniform polynomial-size circuits. In: MCU: Machines, Computations and Universality (accepted, 2013)

29. Neary, T., Woods, D.: P-completeness of cellular automaton rule 110. In: Bugliesi, M., Preneel, B., Sassone, V., Wegener, I. (eds.) ICALP 2006. LNCS, vol. 4051, pp. 132–143. Springer, Heidelberg (2006)

30. Padilla, J., Liu, W., Seeman, N.: Hierarchical self assembly of patterns from the Robinson tilings: DNA tile design in an enhanced tile assembly model. Natural Computing, 1–16 (2011)

31. Padilla, J., Patitz, M., Pena, R., Schweller, R., Seeman, N., Sheline, R., Summers, S., Zhong, X.: Asynchronous signal passing for tile self-assembly: Fuel efficient computation and efficient assembly of shapes. In: Mauri, G., Dennunzio, A., Manzoni, L., Porreca, A.E. (eds.) UCNC 2013. LNCS, vol. 7956, pp. 174–185. Springer, Heidelberg (2013)

32. Papadimitriou, C.M.: Computational complexity, 1st edn. Addison-Wesley Publishing Company, Inc. (1994)

33. Patitz, M.J.: An introduction to tile-based self-assembly. In: Durand-Lose, J., Jonoska, N. (eds.) UCNC 2012. LNCS, vol. 7445, pp. 34–62. Springer, Heidelberg (2012)

34. Prusinkiewicz, P., Lindenmayer, A.: The algorithmic beauty of plants. Springer (1990)

35. Reif, J., Slee, S.: Optimal kinodynamic motion planning for 2D reconfiguration of self-reconfigurable robots. Robot. Sci. Syst. (2007)

36. Rothemund, P.W.K., Winfree, E.: The program-size complexity of self-assembled squares (extended abstract). In: STOC: Proceedings of the Thirty-Second Annual ACM Symposium on Theory of Computing, pp. 459–468. ACM Press (2000)

37. Rus, D., Vona, M.: Crystalline robots: Self-reconfiguration with compressible unit modules. Autonomous Robots 10(1), 107–124 (2001)

38. Soloveichik, D., Cook, M., Winfree, E., Bruck, J.: Computation with finite stochastic chemical reaction networks. Natural Computing 7(4), 615–633 (2008)

39. Summers, S.: Reducing tile complexity for the self-assembly of scaled shapes through temperature programming. Algorithmica, 1–20 (2012)

40. Vollmer, H.: Introduction to Circuit Complexity: A Uniform Approach. Springer-Verlag New York, Inc. (1999)

41. Winfree, E.: Algorithmic Self-Assembly of DNA. PhD thesis, California Institute of Technology (June 1998)

42. Woods, D.: Upper bounds on the computational power of an optical model of computation. In: Deng, X., Du, D.-Z. (eds.) ISAAC 2005. LNCS, vol. 3827, pp. 777–788. Springer, Heidelberg (2005)

43. Woods, D., Chen, H.-L., Goodfriend, S., Dabby, N., Winfree, E., Yin, P.: Active self-assembly of algorithmic shapes and patterns in polylogarithmic time. In: ITCS 2013: Proceedings of the 4th Conference on Innovations in Theoretical Computer Science, pp. 353–354. ACM (2013), Full version: arXiv:1301.2626 [cs.DS]

44. Woods, D., Naughton, T.J.: Parallel and sequential optical computing. In: Dolev, S., Haist, T., Oltean, M. (eds.) OSC 2008. LNCS, vol. 5172, pp. 70–86. Springer, Heidelberg (2008)

45. Yurke, B., Turberfield, A.J., Mills Jr., A.P., Simmel, F.C., Nuemann, J.L.: A DNA-fuelled molecular machine made of DNA. Nature 406, 605–608 (2000)

46. Lvarez, C., Jenner, B.: A very hard log-space counting class. Theoretical Computer Science 107(1), 3–30 (1993)

DNA Walker Circuits: Computational Potential, Design, and Verification

Frits Dannenberg[1], Marta Kwiatkowska[1],
Chris Thachuk[1], and Andrew J. Turberfield[2]

[1] University of Oxford, Department of Computer Science, Wolfson Building, Parks Road,
Oxford, OX1 3QD, UK
[2] University of Oxford, Department of Physics, Clarendon Laboratory, Parks Road,
Oxford, OX1 3PU, UK

Abstract. Unlike their traditional, silicon counterparts, DNA computers have natural interfaces with both chemical and biological systems. These can be used for a number of applications, including the precise arrangement of matter at the nanoscale and the creation of smart biosensors. Like silicon circuits, DNA strand displacement systems (DSD) can evaluate non-trivial functions. However, these systems can be slow and are susceptible to errors. It has been suggested that localised hybridization reactions could overcome some of these challenges. Localised reactions occur in DNA 'walker' systems which were recently shown to be capable of navigating a programmable track tethered to an origami tile. We investigate the computational potential of these systems for evaluating Boolean functions. DNA walkers, like DSDs, are also susceptible to errors. We develop a discrete stochastic model of DNA walker 'circuits' based on experimental data, and demonstrate the merit of using probabilistic model checking techniques to analyse their reliability, performance and correctness.

1 Introduction

The development of simple biomolecular computers is attractive for engineering and health applications that require *in vitro* or *in vivo* information processing capabilities. DNA computing models which use hybridization and strand displacement reactions to perform computation have been particularly successful. DNA strand displacement systems (DSD) have been shown experimentally to simulate logic circuits [12, 13] and are known to be Turing-universal [11]. However, computing with biomolecules creates many challenges. For example, reactions within a DSD are global in the following sense: strands which are intended to react must first encounter one another in a mixed solution. The mixing of all reactants may lead to unintended reactions between strands. These systems do not, at present, ensure the spatial locality typical of other computing models. Qian and Winfree suggested that tethering DNA based circuits to an origami tile could overcome some of these challenges [12]. This idea was explored and expanded upon by Chandran *et al.* [5], who investigate how such systems could be realised experimentally, give constructions of composable circuits, and propose a biophysical model for verification of tethered, hybridization-based circuits. Our work is largely inspired by theirs, but we consider another setting which also exhibits localised reactions: DNA walker systems [2, 7, 10, 14–16].

D. Soloveichik and B. Yurke (Eds.): DNA 2013, LNCS 8141, pp. 31–45, 2013.
© Springer International Publishing Switzerland 2013

Stepping mechanics

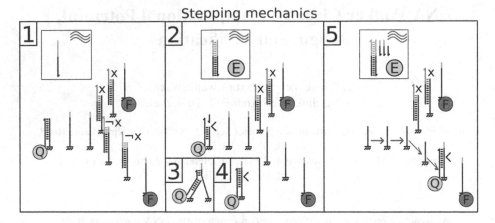

Fig. 1. (1) The walker strand carries a load (Q) that will quench fluorophores (F) when nearby. The walker is attached to the initial anchorage and all other anchorages are blocked. By adding unblocking strands, the selected track becomes unblocked. In this case the signal that opens up the path labelled by $\neg X$ is added. (2) The nicking enzyme (E) attaches to the walker-anchorage complex, and cuts the anchorage. The anchorage top melts away from the walker, exposing 6 nucleotides as a toehold. (3) The exposed toehold becomes attached to the next anchorage. (4) In a displacement reaction, the walker becomes completely attached to the new anchorage. The stepping is energetically favourable, because it re-forms the base pairs that were lost after the previous anchorage was cut. (5) Repeating this process, the walker arrives at a junction. The walker continues down the unblocked track, eventually reaching the final anchorage and quenching the fluorophore.

Various DNA walkers have been experimentally realised — see [14] and references therein. Single-legged DNA walkers were recently shown capable of navigating a programmable track of strands, called *anchorages*, that are tethered to a DNA origami tile [14]. Movement of the walker between anchorages is shown in Fig. 1. Initially, all tracks are blocked by hybridization to blocker strands. Anchorages and their blockers are addressed by means of distinct toehold sequences (shown coloured): anchorages are selectively unblocked by adding strands complementary to their blockers as *input*. Much like field programmable gate arrays, these systems are easily reconfigured. By using programmable anchorages at track junctions, Wickham *et al.* [14] demonstrate that a walker can be directed to any leaf in a complete two-level binary tree using input strands that unblock the intended path.

In Section 2, the computational expressiveness of such walker systems is explored, using a theoretical framework that assumes ideal conditions. We highlight significant limitations of current walker systems and motivate future work. In Section 3 we develop a probabilistic model to analyse the impact of different sources of error that arise in experiments on reliability, performance and correctness of the computation. The model can be used to support the design and verification of DNA walker *circuits*.

2   Computational Potential of DNA Walker Circuits

In this section we explore the computational potential of DNA walker systems. We focus on deterministic Boolean function evaluation and call the resulting constructions *DNA walker circuits*. We begin by defining a model of computation that makes explicit the underlying assumptions that characterize the DNA walker systems considered here. These assumptions are consistent with current published experimental systems: in particular, we do not explore the potential for multiple walkers to interact within the same circuit. However, we do consider the potential consequences for parallel computation.

2.1   A Model of Computation for DNA Walker Circuits

A *DNA walker circuit* is composed of straight, undirected, *tracks* (consecutive anchorages), and *gates* (track junction points) that connect at most three tracks. A gate can have at most one Boolean *guard* for each track that it connects. A particular guard is implemented using one or more blocking strands that share a common toehold sequence; distinct guards use distinct toehold sequences. A track adjacent to a gate is *blocked* if it has a guard that evaluates to false — its unblocking strands are not added to solution — and is unblocked otherwise. For example, Fig. 1 depicts a circuit of a single gate connecting three tracks. The track ending with the anchorage marked with the red fluorophore (top right of panel 1) has the Boolean guard X, while the track ending with the anchorage marked with the green fluorophore has the Boolean guard $\neg X$. Panel 2 of Fig. 1 shows that the path to the green fluorophore is unblocked when $\neg X$ evaluates to true (*i.e.*, the unblocking strands for $\neg X$ are added to solution). In this case, X evaluates to false and the path to the red fluorophore remains blocked (*i.e.*, the unblocking strands for X are *not* added to solution). We define a *fork gate* as having at most one input track, and exactly two guarded output tracks. Each circuit has one *source* – a fork gate with no input track denoting the initial position of a walker. A *join gate* with an output track has at most two guarded input tracks. A join gate with no output track is a *sink* and has at most three (unguarded) input tracks. Each circuit has one or more *true sinks* and one or more *false sinks*.

In a circuit C with Boolean guards over n variables, a *variable assignment* A for C is a truth assignment of those n variables. Consider any DNA walker circuit C and variable assignment A for C. Let $C[A]$ denote the set of reachable paths originating from the source of C, after all guards are evaluated as blocked or unblocked, under assignment A. We say that C is *deterministic* under assignment A if there is exactly one path from the source to a sink in $C[A]$. Note that this definition of determinism precludes the possibility of a *deadlock*, (*i.e.*, when no path from the source can reach a sink). Let VALUE $(C[A])$ be the *output value* of the circuit under assignment A (*i.e.*, whether the reachable sink is a true sink or a false sink). Circuit C is *deterministic* if it is deterministic under all possible variable assignments.

A *circuit set* S, consisting of one or more unconnected circuits, is deterministic if and only if VALUE $(C_i[A])$ = VALUE $(C_j[A])$, for each $C_i, C_j \in$ S, under any possible assignment A. Let VALUE $(S[A])$ be the value of S under assignment A. The *size* of S, denoted by SIZE (S), is the total count of component gates.[1] We define the worst

---

[1] We do not investigate circuit *area* in this paper.

case *time* of a computation in S, denoted by TIME (S), as the longest reachable path from a source to a sink. This notion of time captures the ability of multiple walkers to simultaneously traverse disjoint paths (one per unconnected circuit).

Let $S[A]$ denote the set of reachable paths in S under assignment A (one per unconnected circuit). Given a circuit $C_i \in S$, we say that a gate $G \in C_i$ is *reachable* in $C_i[A]$ (equivalently $S[A]$) if there exists an unblocked path from the source of C_i to G. Note that, if every gate is reachable, this implies that every output track of a gate can be traversed under some variable assignment. We call gates where this is not true *redundant*. We will reason about circuit sets where all gates are reachable and non-redundant under some variable assignment. When this is not the case, the circuit set can be simplified to one that is logically equivalent.

2.2   Reporting Output in DNA Walker Circuits

Output of a DNA walker circuit can be reported with the use of different coloured (spectrally resolvable) fluorophores and also quenchers. If a walker carries a quencher cargo, then it has the potential to decrease one of a number of different fluorescent signals from fluorophores positioned at the circuit sinks. This scenario is illustrated in Fig. 2 (Left). In a circuit that decides a Boolean function, a single, quenching, walker can only decrease the signal of a particular colour (corresponding to a particular fluorophore) by an amount that is inversely proportional to the number of sinks labelled with that same colour. Accurate output reporting could be problematic in larger circuits

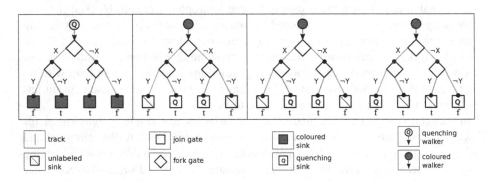

Fig. 2. Reporting Boolean decisions with DNA walker circuits. (Left) A quenching walker with red fluorophores labelling false sinks and green fluorophores labelling true sinks. A drop in signal for one colour indicates the truth value of the circuit. However, the signal drop is inversely proportional to the number of sinks of the same colour. (Center) A green coloured walker and quenching true sinks. When the circuit evaluates to true the green signal is fully suppressed. However, the fluorescence output from this circuit cannot distinguish between an incomplete computation and a false one. (Right) Two parallel copies of the circuit, with different fluorophores labelling the walkers and with quenching true sinks in one and quenching false sinks in the other: the computation is complete and unambiguously reported when one colour is suppressed.

with many sinks. We will therefore focus only on reporting strategies that fully suppress a particular colour. Rather than carrying a quencher, a walker instead carries a fluorophore of a single colour and either all true sinks or all false sinks are labelled with quenchers. An example with quenching true sinks is shown in Fig. 2 (Center). This circuit can fully suppress the fluorophore signal when it evaluates to true, regardless of its size. However, this is a one-sided reporting strategy as one cannot distinguish between the case of an incomplete computation or one evaluating to false. As illustrated in Fig. 2 (Right) this shortcoming can be addressed by using two circuits in parallel with each using a one-sided reporting strategy. Each of the two (otherwise identical) circuits uses a different coloured walker: one has quenching false sinks and the other quenching true sinks. In this circuit set, one colour will be fully suppressed when it is true, the other when it is false, and neither will be suppressed until the computation completes.

2.3 Deterministic Fork and Join Gates in DNA Walker Circuits

If all gates in a circuit set S are deterministic, it follows that S is deterministic. The following theorem shows that deterministic fork gates must have output guards that are negations of each other.

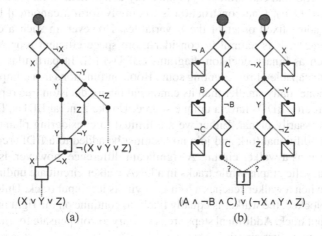

Fig. 3. (a) A connectivity graph of a DNA walker circuit to evaluate the disjunction $(X \vee Y \vee Z)$. There are two output tracks: one when the circuit evaluates to true, the other when it evaluates to false. The resulting path when $X = Y = f$ and $Z = t$ is shown highlighted. (b) Two conjunction circuits are composed into the disjunction $(A \wedge \neg B \wedge C) \vee (\neg X \wedge Y \wedge Z)$. Two source nodes (two walkers) are used to evaluate clauses in parallel. No assignment of guards to the join gate labelled J can ensure that this circuit is deterministic. This is evident when $A = C = Y = Z = t$ and $B = X = f$.

Theorem 1. *A fork gate in a DNA walker circuit is deterministic if and only if there exists some guard G such that the left output track is guarded by G and the right is guarded by ¬G.*

Proof. If neither output track is guarded, then any path that can reach the gate could be extended via the left or the right output track and the gate would not be deterministic. Similarly, this is true when only one output track is guarded and the guard evaluates to false. (If the fork gate is only reachable when the single output guard evaluates to true, then the gate is redundant as the output track with the guard is never used.) Thus, consider when each output track is guarded and let the left output have guard G_L and the right have guard G_R. Note that $G_L \not\equiv G_R$ as otherwise any path that reaches the gate will result either in a deadlock — when both evaluate to false — or the path could be extended via the left or the right output tracks — when both evaluate to true. Consider any path p that can reach the gate and the case when G_L evaluates to true and G_R to false. It follows that, before reaching the gate, p must not traverse a track guarded by $\neg G_L$ nor by G_R. Since the gate is non-redundant, p must also be able to reach the gate when G_L evaluates to false and G_R to true. It follows that, before reaching the gate, p must not traverse a track guarded by G_L nor by $\neg G_R$. Therefore, path p is independent of the variables affecting guards G_L and G_R. Thus, there exists a variable assignment such that any path reaching the gate will result in a deadlock, or can be extended via both output tracks, unless $G_L \equiv \neg G_R$. □

Given any Boolean function $f : \{0, 1\}^n \to \{0, 1\}$, there exists a deterministic DNA walker circuit set S that can evaluate f, under any assignment to its n variables, such that TIME (S) $= O(n)$. One construction is to simply form a canonical binary decision tree over some fixed order of the n variables. However, in such a construction SIZE (S) $= \Theta(2^n)$. It is natural to consider more space efficient representations to evaluate f, such as binary decision diagrams (BDDs) [4]. In particular, reduced ordered BDDs are capable of representing some Boolean functions in a compressed form that can be exponentially smaller than its canonical binary decision tree representation. Like walker circuits, BDDs have a unique source. Unlike general BDDs, DNA walker circuits are necessarily planar. Either we are limited to considering planar BDD representations or additional fork and join nodes must be added to a BDD representation when realising it as a walker circuit. A significant difference, however, is that BDDs form directed acyclic graphs while tracks in a DNA walker circuit are undirected. Consider the case when a walker reaches a join gate via its left input track. Unless the right input track is blocked, the walker is equally likely to continue on the right input track as it is on the output track. Additional steps are necessary to compensate for the undirected nature of tracks in walker circuits.

Unlike fork gates, it is not obvious whether all join gates can be made deterministic. Theorem 2 characterizes both the necessary and sufficient conditions: a deterministic join of two disjoint sets of paths, one for each input track, is only possible if they were previously "forked"[2] on some variable X (*i.e.*, in one set all paths traverse an edge guarded by X and in the other set all traverse an edge guarded by $\neg X$). This property is exemplified by the contrast between the disjunction circuit of Fig. 3(a) and the disjunction of two conjunctions circuit as shown in Fig. 3(b). In the latter, two walkers are used in an attempt to parallelize the evaluation. However, as the clauses do not have literals over a common variable, there are no guards that can be assigned to

---

[2] It is not a necessary condition that the two disjoint sets of paths reaching the join were forked by a common gate, only that they can be partitioned based on the value of some variable.

the join gate labeled J to ensure the circuit is deterministic. Note that this limitation is not caused by the restricted topology of walker circuits (*i.e.*, their layout on a planar surface), but rather by the property that their tracks are undirected.

Theorem 2. *A join gate in a DNA walker circuit is deterministic if and only if there exists some guard G such that the left input track is guarded by G, the right by ¬G and, prior to reaching those guards, all paths that can reach the left input must traverse a track guarded by G and all paths that can reach the right must traverse a track guarded by ¬G.*

Proof. (⇒ *if*) Suppose the left input track is guarded by G, the right by ¬G and, prior to reaching those guards, all paths that can reach the left input must traverse a track guarded by G and all paths that can reach the right must traverse a track guarded by ¬G. There are two cases to consider. Suppose G evaluates to true. Then, no path can reach the right input since, by the assumption, those paths must traverse a track guarded by ¬G prior to reaching the gate. It follows that all paths that can reach the gate when G evaluates to true must be to the left input. Furthermore, as the right input is guarded by ¬G, those paths can only be extended via the output of the gate. The other case (G evaluates to false) is symmetric. Furthermore, as the guards are negations of each other, they cannot simultaneously evaluate to false and cause a potential deadlock.

(⇐ *only if*) Let G_L and G_R be the guards of the left and right inputs, respectively. (If one or more of the input tracks is unguarded, then the gate cannot be deterministic when both are reachable by at least one path.) First, consider all paths that can reach the left input, guarded by G_L. It must simultaneously be true that none of those paths (i) traverse a track guarded by ¬G_L and (ii) all of those paths traverse a track guarded by ¬G_R. If condition (i) is not satisfied, then there would exist a path that traverses a track guarded by ¬G_L and, to extend past the join gate, must traverse another guarded by G_L. As this is not possible, the path would end in a deadlock and the gate would not be deterministic. If condition (ii) is not satisfied then there would exist some path p that does not traverse a track guarded by ¬G_R, but may possibly traverse a track guarded by G_R. In this case, there exists a variable assignment where G_R, and all other guards on path p, evaluate to true. With such a variable assignment, path p could be extended via the output track or the right input track. Thus, condition (ii) must also be satisfied, as otherwise the gate would not be deterministic. The conditions (and the argument that both are necessary) when considering all paths that can initially reach the right input, guarded by G_R, are symmetric.

The sufficiency argument (⇒ *if*) shows the gate is deterministic when $G_L \equiv \neg G_R$. It remains to show it is not deterministic otherwise. First, consider the consequence when both G_L and G_R evaluate to true. By condition (ii) all paths leading to the left (right) input traverse a track guarded by ¬G_R (¬G_L). In this case, no paths can reach the gate. Recall that the gate is non-trivial and therefore each input is reachable by at least one path. Thus, consider when both G_L and G_R evaluate to false. The conditions permit that paths can reach the gate; however, if any path does it will deadlock as both inputs to the gate are blocked. Thus, for all paths that can reach the gate, it will be deterministic only when $G_L \equiv \neg G_R$.   □

2.4   Evaluating Boolean Formulas with DNA Walker Circuits

Despite the shortcomings of join gates in current DNA walker circuits, it is not the case that Boolean formulas must be evaluated using a circuit forming a binary decision tree. Any Boolean formula can be represented in one of its canonical forms. We will focus on conjunctive normal form (CNF) which is a single conjunction of clauses, where each clause is a disjunction over literals. A formula in CNF is said to be k-CNF if the largest clause has size k. Using a standard transformation, a Boolean formula in k-CNF with at most l total literals can be converted to a 3-CNF formula over $O(l)$ variables, with at most $O(l)$ clauses (each having at most 3 literals). As such, we will reason exclusively about circuits to evaluate 3-CNF formulas.

Constructing a walker circuit to represent a formula in 3-CNF with m clauses is straightforward. Each clause can be represented by the disjunction circuit of Fig. 3(a). The source of the circuit will be the first fork gate of the first clause. The output track signalling the i-th clause is satisfied is connected to the input track of clause $i + 1$. Thus, the walker will only reach the single true sink of the circuit (output from clause m) if the formula is satisfied for that particular variable assignment. To ensure that both true and false signals can be reported deterministically, we use the reporting strategy depicted in Fig. 2 (Right) which employs two parallel copies of the circuit, each using different coloured walkers and different quenching sinks.

Theorem 3. *Let $\mathcal{F}$ be any 3-CNF Boolean formula with m clauses. There exists a DNA walker circuit set S, with* SIZE(S) $= \Theta(m)$ *and* TIME(S) $= O(m)$, *such that given any variable assignment A for $\mathcal{F}$,* VALUE(S$[A]$) *is the truth value of $\mathcal{F}$ under assignment A.*

Proof. The construction is described in Section 2.4 and it is easy to see that the circuit is deterministic and that it correctly reports the truth value of $\mathcal{F}$ under assignment A. What remains is to bound the circuit size and worst case time. The construction uses a set of two circuits: $\mathcal{C}_T$ and $\mathcal{C}_F$. Consider the circuit $\mathcal{C}_T$ used to evaluate if $\mathcal{F}$ is true under assignment A. There are m clauses and each is simulated by a disjunction circuit of size $O(1)$. These circuits are composed in series to form $\mathcal{C}_T$. Therefore, SIZE($\mathcal{C}_T$) $= \Theta(m)$ and TIME($\mathcal{C}_T$) $= O(m)$. The arguments are the same for circuit $\mathcal{C}_F$ and, as both are evaluated in parallel, the claim follows.   □

While the construction of Theorem 3 can represent any Boolean formula, and some in exponentially less space than a binary decision tree, the resulting circuit set is formula specific. Given the effort of creating DNA walker circuits, a more uniform circuit — one capable of evaluating many Boolean functions — is worth exploring. As with silicon circuits, we can construct a uniform circuit to evaluate any 3-CNF formula, under any variable assignment, up to some bound on the number of variables. Each variable can be present in a clause as either a positive or negative literal, but not both. (The circuit can be modified to handle this case if necessary.) Therefore, there are at most $2^3 \binom{n}{3}$ unique clauses in any 3-CNF Boolean formula over n variables, and also for any formula over $m \leq n$ variables. In this general circuit, we supplement each possible clause with an initial fork gate guarded on the condition of the clause being active or inactive in the particular formula being evaluated. If it is inactive, the walker can pass through to the output track denoting true, without traversing guards for the literals of the clause. Note that this only increases the size of each clause by a constant.

Corollary 1. *There exists a DNA walker circuit set* S, *with* SIZE(S) $= O(n^3)$ *and* TIME(S) $= O(n^3)$, *that can evaluate any 3-CNF Boolean formula over* $m \leq n$ *variables under any variable assignment.*

A 3-CNF formula with m clauses can be evaluated in polylogarithmic time (in m) using a silicon circuit in a straightforward manner: each clause can be evaluated in parallel and those results can be combined using a binary reduction tree of height $O(\log m)$— only if all clauses are satisfied will the root of the reduction tree output true. Is the same possible in DNA walker circuits? Unfortunately, this is not the case in general. Such a circuit would require a new kind of join gate, outside of our current model of computation, to perform a conjunction of multiple walkers — one walker leaves the gate only after all input walkers have arrived. Parallel evaluation of circuits representing formulas in disjunctive normal form (DNF) does not fair better. Consider the case of a DNF formula with m clauses where clause $m - 1$ and clause m have no literals over a common variable. By Theorem 2, a join gate connecting the circuits for these clauses cannot be deterministic. An example of this situation is given in Fig. 3(b).

3 Design and Verification of DNA Walker Circuits

We have so far assumed DNA walker circuits to work perfectly. In a real experiment various errors can occur, for example, the walker may release from the track, or a blockade can fail to block an anchorage. In this section, we analyse the reliability and performance of DNA walker circuits using probabilistic model checking. We develop a continuous-time Markov chain model, based on a variety of DNA walker experiments from [2, 14, 15], and analyse it against quantitative properties such as the probability of the computation terminating or expected number of steps taken until termination. We use the PRISM model checker [8], which accepts models described in a scripted language and properties in the form of temporal logic. For example, if we label all states of the model where a walker quenches any fluorophore by "finished", then the query $P_{=?}[\,F^{[T,T]}\,\text{finished}\,]$ yields the probability of all paths that eventually reach a state where a walker has quenched a fluorophore (in other words, the computation terminated) by time T. A custom tool was developed to generate PRISM model scripts with matching track-layout graphs. Different configurations of tracks are studied: linear tracks are considered in Fig. 4 (Top), while branched tracks are used in Fig. 5 and Fig. 6. We use the results of experiments on linear (Fig. 4) and single-branched tracks to establish model parameters, and match model predictions with observations on double-layer tracks to evaluate the quality of our model.

Experiments show that the walker can step onto anchorages that are fixed as far away as 19 nm. We assume non-zero rates for the walker to step onto *any* intact anchorage within 24 nm distance. This range was chosen by taking into account the lengths of the empty anchorage and walker-anchorage complex, estimated around 15 nm and 11 nm respectively.

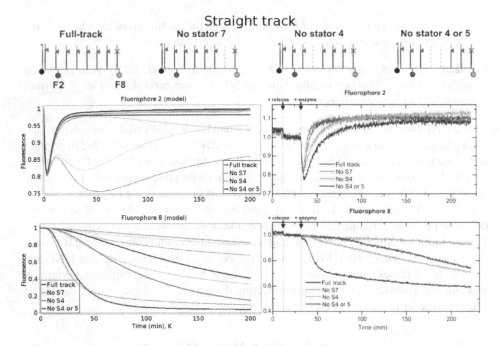

Fig. 4. Top: A small linear track of 8 anchorages with fluorophores on both the second and last anchorage. Experiments were performed with one or more anchorages omitted [15]. Right: Experimental results (reproduced with permission from the authors). The walker hardly reaches the final anchorage when anchorage 7 is removed, due to the double penalty of a longer final step *and* the mismatch in the final anchorage. Left: Model results. Dotted lines: Alternative model where the walker can step onto already-cut anchorages with rate $k_b = k_s/30$.

A step taken by the walker corresponds to a single transition in the Markov chain, although the real stepping process is more complex, as depicted in Fig. 1. Assume that the *stepping rate* k depends on distance d between anchorages and some *base stepping rate* k_s. Denote by $d_a = 6.2$ nm the average distance between anchorages in the experiment shown in Fig. 4. Denote by $d_M = 24$ nm the *maximal interaction* distance discussed earlier. Based on previous experimental estimates of [15], we fix the stepping rate k as:

$$k = \begin{cases} k_s = 0.009\text{s}^{-1} & \text{when } d \leq 1.5d_a \\ k_s/50 & \text{when } 1.5d_a < d \leq 2.5d_a \\ k_s/100 & \text{when } 2.5d_a < d \leq d_M \\ 0 & \text{otherwise} \end{cases} \quad (1)$$

These rates define a sphere of reach around the walker-anchorage complex, allowing the walker to step onto an uncut anchorage when it is nearby. In Fig. 5 the sphere of reach is depicted to scale with walker circuits. There are two exceptions. Stepping *from* the initial anchorage and stepping *onto* the final anchorage occur at lower rates.

The domain complementary to the walker on the initial anchorage is two bases longer than the corresponding domain of a regular anchorage. Stepping from the initial anchorage was reported to happen $3\times$ more slowly: this is incorporated in the model. The final anchorage includes a mismatched base that prevents cutting by the nicking enzyme. Based on the experimental data, we fit a tenfold reduction for the rate of stepping *onto* the final absorbing anchorage (Fig. 4).

Three types of interaction that are known to occur are omitted from the model: all three could be incorporated in future. Firstly, a rate of $k_s/5000$ is reported [15] for transfer of the walker between *separate* tracks built on different DNA origami tiles. Transfer between tracks could be eliminated by binding the tiles to a surface, thus keeping them apart. Secondly, the walker can move between *intact* anchorages in the absence of the nicking enzyme with a rate of $\sim k_s/13$ [15]. With the enzyme present, the walker spends little time attached to an intact anchorage, as enzymatic activity is relatively fast.[3] Therefore we remove the rate altogether. In our model, the anchorage is cut as soon as the walker attaches to it. Thirdly, the walker can step backward onto cut anchorages. This requires a blunt-end strand-displacement reaction which is known to be slow relative to toehold-mediated displacement [17]. A variant of the model with a *backward* rate $k = k_s/30$ is shown in dotted lines in Fig. 4 (Left). In this case the model predicts significant quenching of fluorophore F2 at late times by walkers whose forward motion is obstructed by omission of one or more anchorages: this does not match experimental data. A reduced rate $k_b = k/500$ (not plotted) has a similar effect.

The time-dependent responses of fluorescent probes F2 and F8 shown in Fig. 4 (Left) are predicted by the Markov chain model using the rate parameters discussed above without any further fitting: they correspond well to the experimental data.

An additional parameter is needed to model branched tracks (Fig. 5(a)). We introduce a failure rate for the anchorage blocking mechanism which is assumed to be the same for all junctions. We infer a failure rate of 30% by fitting to the results of the single-layer branched-track experiment illustrated in Fig. 5 [14].

3.1 Model Results

Having used experiments on straight tracks and with a single layer of branching to determine the parameters of the model, we use the two-layer junction experiments shown in Fig. 5(c) to evaluate its quality. The model captures essential features of the walker behaviour and is reasonably well aligned with experimental data. In the model, not all walkers reach an absorbing anchorage by time $T = 200\text{min}$, although the predicted quenching is much higher than observed. The reason for this discrepancy is not easily determined and motivates further study.

We exercise the model by model checking them against temporal logic queries aimed at quantifying the reliability and performance of the computation. We note that not all the walkers that finish actually do quench the intended signal. In both the model and the experiments we can identify a difference between paths that follow the side of the track (paths LL and RR), and paths that enter the interior (paths RL and LR):

---

[3] The cutting rate for enzymatic activity was measured at $0.17s^{-1}$, for which the enzyme binding to the DNA is considered not a rate limiting step [3].

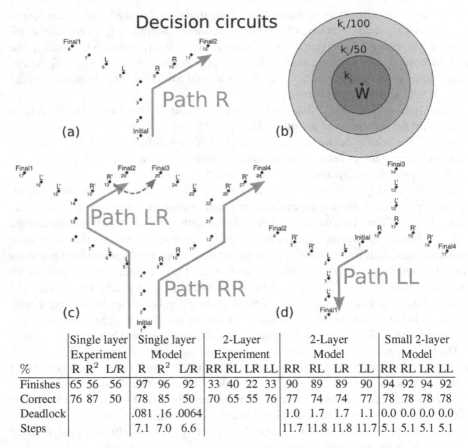

%	Single layer Experiment			Single layer Model			2-Layer Experiment				2-Layer Model				Small 2-layer Model			
	R	R^2	L/R	R	R^2	L/R	RR	RL	LR	LL	RR	RL	LR	LL	RR	RL	LR	LL
Finishes	65	56	56	97	96	92	33	40	22	33	90	89	89	90	94	92	94	92
Correct	76	87	50	78	85	50	70	65	55	76	77	74	74	77	78	78	78	78
Deadlock				.081	.16	.0064					1.0	1.7	1.7	1.1	0.0	0.0	0.0	0.0
Steps				7.1	7.0	6.6					11.7	11.8	11.8	11.7	5.1	5.1	5.1	5.1

Fig. 5. Top: Track topology for single-layer (a) and double-layer (c,d) decision tracks. Initial indicates the initial anchorage, Final indicates absorbing anchorages, and L, L', R and R' indicate anchorages that can be blocked by input. Coloured circles (b) indicate the range of interaction of the walker to scale. Bottom: Experimental results [14] compared with results from the model. Single layer track: R means a single blockade on the left, R^2 means a two-anchorage blockade on the left, L/R means single blockades on both the left and right. Double layer track: RL means anchorages labelled L and R' are blocked, so that the walker goes right on the first decision, and left on the second. Each blockade is of two consecutive anchorages. All properties are given at time $T = 200$ min. *Finishes*, $P_{=?}[F^{[T,T]}$ finished], is the probability that a walker quenches any fluorophore by time T; *Correct*, $P_{=?}[F^{[T,T]}$ ("finished-correct"|"finished")], is the probability that a finished walker quenches the correct fluorophore by time T; *Deadlock*, $P_{=?}[F^{[T,T]}$ deadlock], is the probability for the walker to get stuck prematurely by time T, with no intact anchorage within reach; and *Steps*, $R_{=?}$ (steps) $[C^{\leq T}]$, indicates the expected number of steps taken by time T.

the probability of a correct outcome for the side paths is greater. This is explained by *leakage transitions* between neighbouring paths, for example, see the red dotted line in Fig. 5(d). Walkers on an interior path can leak to both sides, but a path that follows the side can only leak to one side. This effect can also be shown by inspecting paths. By using the property $P_{=?}[$ correct-path $U^{\leq T}$ finished-correct], which denotes

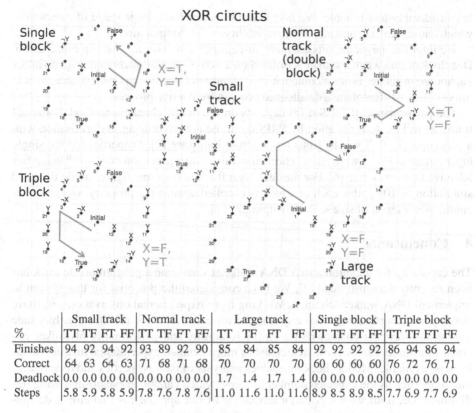

Fig. 6. Performance analysis for a logic track expressing the XOR formula $(X \oplus Y)$. Properties as in Fig. 5.

%	Small track				Normal track				Large track				Single block				Triple block			
	TT	TF	FT	FF	TT	TF	FT	FF	TT	TF	FT	FF	TT	TF	FT	FF	TT	TF	FT	FF
Finishes	94	92	94	92	93	89	92	90	85	84	85	84	92	92	92	92	86	94	86	94
Correct	64	63	64	63	71	68	71	68	70	70	70	70	60	60	60	60	76	72	76	71
Deadlock	0.0	0.0	0.0	0.0	0.0	0.0	0.0	0.0	1.7	1.4	1.7	1.4	0.0	0.0	0.0	0.0	0.0	0.0	0.0	0.0
Steps	5.8	5.9	5.8	5.9	7.8	7.6	7.8	7.6	11.0	11.6	11.0	11.6	8.9	8.5	8.9	8.5	7.7	6.9	7.7	6.9

the probability that a walker stays on the path until it quenches the correct fluorophore by time T, we can reason about the likelihood of the walker deviating from the intended path. For the double-layer track in Fig. 5(d), we infer that the probability of staying on the intended path *and* reaching the absorbing anchorage within 200 minutes is 55% for paths LR and RL, and 58% for paths LL and RR. This shows that walkers on interior paths are indeed more likely to deviate from the intended path than walkers on paths that follow the sides.

The double-layer track can be optimized by reducing the probability of leakage from the intended path. By decreasing the proximity of off-path anchorages and reducing the track length, both the proportion of walkers finishing and correctness are increased (see Fig. 5(d)). The asymmetry between paths (LL, RR vs. LR, RL) also disappears.

Smaller tracks are not always better. In Fig. 6 several variants of a XOR-logic circuit are shown. The 'small', 'normal' and 'large' variants all use a total of four blocker strands per decision node. The large track is approximately as correct as the normal sized track, but a lower proportion of walkers reach an absorbing anchorage. The small track has a greater proportion of walkers that finish than the normal sized track, but it is considerably less reliable. We note that the walker has a large range of interaction, which causes leakage and affects the reliability of the computation.

We infer that larger circuits are more susceptible to *deadlock*, based on Fig. 5 and 6. Deadlock occurs when a walker is isolated on a non-absorbing anchorage with no intact anchorage in range. From a computational standpoint deadlock is undesirable, as it is impossible to differentiate a deadlocked process from a live process.

The performance of PRISM [8] depends on the model checking method. For small tracks as in Fig. 4, verification by PRISM can be achieved using uniformisation with a precision of 10^{-6} within 10ms on common hardware [1]. Properties for the single layer circuit in Fig. 5 were model checked within 3s to a precision of $< 10^{-6}$ using fast adaptive uniformisation [6]. For the dual-layer track in the same figure, single-threaded simulation of 10^5 paths, each of which is checked against the property, yields a 95% confidence interval of size $< 0.4\%$ within 23s [1].

4 Conclusions

The capability for an autonomous DNA walker to navigate a programmable track has been recently demonstrated [14]. We have considered the potential for this system to implement DNA walker 'circuits'. Working from experimental observations, we have developed a simple model that explains the influence of track architecture, blockade failure and stepping characteristics on the reliability and performance of walker circuits. The model can be further extended as more detailed experimental measurements become available. Model checking enables analysis of path properties and quantitative measures such as the expected number of steps, which cannot be established using traditional ODE frameworks. A major advantage of our approach is that circuit designs can be manipulated to study the properties of variant architectures.

We have shown that walker circuits can be designed to evaluate any Boolean function. In the experimental system we have considered, paths within a circuit can only be joined under specific conditions, resulting in a number of theoretical consequences. One motivation for implementing circuits with a DNA walker system, instead of a DNA strand displacement system (DSD), is the potential for faster reaction times due to spatial locality. However, the walker system we have considered has severely limited potential for parallel circuit evaluation using multiple walkers. As this is not an issue in a DSD, it is the case that this walker system requires exponentially more time to compute certain Boolean functions than a corresponding DSD. This is not necessarily true of all walker systems. The problem arises in the system under consideration due to the undirected nature of the tracks that are traversed by a walker.

Another autonomous walker system with directed tracks has been demonstrated [16] and, in principle, could be extended to have programmable (directed) tracks. In addition to implementing circuits that could be evaluated efficiently by many walkers in parallel, such a system could also benefit from well established design techniques to improve overall circuit reliability [9]. Furthermore, current walker technology 'destroys' the track that is traversed. New mechanisms that can either replenish the track, or can avoid 'destroying' it, will lead to reusable circuits. Finally, it would be interesting to explore the information processing capabilities of DNA walkers beyond circuit evaluation and the potential for multiple interacting walkers to exhibit emergent behaviour.

Acknowledgements. We thank Masami Hagiya, Jonathan Bath and Alex Lucas for useful discussions. The authors are supported by a Microsoft Research scholarship (FD), ERC AdG VERIWARE, EPSRC Grant EP/G037930/1, a Royal Society - Wolfson Research Merit Award (AJT), and Oxford Martin School.

References

1. Intel i5-2520M, Fedora 3.8.4-102.fc17.x86_64, OpenJDK RE-1.7, PRISM 4.0.3
2. Bath, J., Green, S.J., Turberfield, A.J.: A free-running DNA motor powered by a nicking enzyme. Angewandte Chemie (International ed. in English) 44(28), 4358–4361 (2005)
3. Bellamy, S.R.W., Milsom, S.E., Scott, D.J., Daniels, L.E., Wilson, G.G., Halford, S.E.: Cleavage of individual DNA strands by the different subunits of the heterodimeric restriction endonuclease BbvCI. Journal of Molecular Biology 348(3), 641–653 (2005)
4. Bryant, R.E.: Symbolic Boolean manipulation with ordered binary-decision diagrams. ACM Computing Surveys 24(3), 293–318 (1992)
5. Chandran, H., Gopalkrishnan, N., Phillips, A., Reif, J.: Localized hybridization circuits. In: Cardelli, L., Shih, W. (eds.) DNA 17. LNCS, vol. 6937, pp. 64–83. Springer, Heidelberg (2011)
6. Dannenberg, F., Hahn, E.M., Kwiatkowska, M.: Computing cumulative rewards using fast adaptive uniformisation. In: Proc. 11th Conference on Computational Methods in Systems Biology (CMSB 2013) (to appear)
7. Green, S., Bath, J., Turberfield, A.: Coordinated chemomechanical cycles: A mechanism for autonomous molecular motion. Physical Review Letters 101(23), 238101 (2008)
8. Kwiatkowska, M., Norman, G., Parker, D.: PRISM 4.0: Verification of probabilistic real-time systems. In: Gopalakrishnan, G., Qadeer, S. (eds.) CAV 2011. LNCS, vol. 6806, pp. 585–591. Springer, Heidelberg (2011)
9. von Neumann, J.: Probabilistic logics and synthesis of reliable organisms from unreliable components. In: Shannon, C., McCarthy, J. (eds.) Automata Studies, pp. 43–98. Princeton University Press (1956)
10. Omabegho, T., Sha, R., Seeman, N.C.: A bipedal DNA Brownian motor with coordinated legs. Science 324(5923), 67–71 (2009)
11. Qian, L., Soloveichik, D., Winfree, E.: Efficient Turing-universal computation with DNA polymers. In: Sakakibara, Y., Mi, Y. (eds.) DNA 16 2010. LNCS, vol. 6518, pp. 123–140. Springer, Heidelberg (2011)
12. Qian, L., Winfree, E.: Scaling up digital circuit computation with DNA strand displacement cascades. Science 332(6034), 1196–1201 (2011)
13. Seelig, G., Soloveichik, D., Zhang, D., Winfree, E.: Enzyme-free nucleic acid logic circuits. Science 314(5805), 1585–1588 (2006)
14. Wickham, S.F.J., Bath, J., Katsuda, Y., Endo, M., Hidaka, K., Sugiyama, H., Turberfield, A.J.: A DNA-based molecular motor that can navigate a network of tracks. Nature Nanotechnology 7(3), 169–173 (2012)
15. Wickham, S.F.J., Endo, M., Katsuda, Y., Hidaka, K., Bath, J., Sugiyama, H., Turberfield, A.J.: Direct observation of stepwise movement of a synthetic molecular transporter. Nature Nanotechnology 6(3), 166–169 (2011)
16. Yin, P., Yan, H., Daniell, X.G., Turberfield, A.J., Reif, J.H.: A unidirectional DNA walker that moves autonomously along a track. Angewandte Chemie International Edition 43(37), 4906–4911 (2004)
17. Zhang, D.Y., Winfree, E.: Control of DNA strand displacement kinetics using toehold exchange. Journal of the American Chemical Society 131(47), 17303–17314 (2009)

Leaderless Deterministic Chemical Reaction Networks

David Doty[1] and Monir Hajiaghayi[2]

[1] California Institute of Technology, Pasadena, California, USA
ddoty@caltech.edu
[2] University of British Columbia, Vancouver, BC, Canada
monirh@cs.ubc.ca

Abstract. This paper answers an open question of Chen, Doty, and Soloveichik [5], who showed that a function $f : \mathbb{N}^k \to \mathbb{N}^l$ is deterministically computable by a stochastic chemical reaction network (CRN) if and only if the graph of f is a semilinear subset of $\mathbb{N}^{k+l}$. That construction crucially used "leaders": the ability to start in an initial configuration with constant but non-zero counts of species other than the k species $X_1, \ldots, X_k$ representing the input to the function f. The authors asked whether deterministic CRNs without a leader retain the same power.

We answer this question affirmatively, showing that every semilinear function is deterministically computable by a CRN whose initial configuration contains only the input species $X_1, \ldots, X_k$, and zero counts of every other species. We show that this CRN completes in expected time $O(n)$, where n is the total number of input molecules. This time bound is slower than the $O(\log^5 n)$ achieved in [5], but faster than the $O(n \log n)$ achieved by the *direct* construction of [5].

1 Introduction

In the last two decades, theoretical and experimental studies in molecular programming have shed light on the problem of integrating logical computation with biological systems. One goal is to re-purpose the *descriptive* language of chemistry and physics, which describes how the natural world works, as a *prescriptive* language of programming, which prescribes how an artificially engineered system *should* work. When the programming goal is the manipulation of individual molecules in a well-mixed solution, the language of chemical reaction networks (CRNs) is an attractive choice. A CRN is a finite set of reactions such as $X + Y \to X + Z$ among abstract molecular species, each describing a rule for transforming reactant molecules into product molecules.

CRNs may model the "amount" of a species as a real number, namely its concentration (average count per unit volume), or as a nonnegative integer (total count in solution, requiring the total volume of the solution to be specified as part of the system). The latter integer counts model is called "stochastic" because reactions that discretely change the state of the system are assumed to happen probabilistically, with reactions whose reactants have high molecular counts more likely to happen first than reactions whose molecular counts are

D. Soloveichik and B. Yurke (Eds.): DNA 2013, LNCS 8141, pp. 46–60, 2013.
© Springer International Publishing Switzerland 2013

smaller. The computational power of CRNs has been investigated with regard to simulating boolean circuits [12], neural networks [10], digital signal processing [11], and simulating bounded-space Turing machines with an arbitrary small, non-zero probability of error with only a polynomial slowdown [3]. CRNs are even efficiently Turing-universal, again with a small, nonzero probability of error over all time [13]. Certain CRN termination and producibility problems are undecidable [8,16], and others are PSPACE-hard [15]. It is also difficult to design a CRN to "delay" the production of a certain species [6,7]. Using a theoretical model of DNA strand displacement, it was shown that any CRN can be transformed into a set of DNA complexes that approximately emulate the CRN [4]. Therefore even hypothetical CRNs may one day be reliably implementable by real chemicals.

While these papers focus on the stochastic behaviour of chemical kinetics, our focus is on CRNs with deterministic guarantees on their behavior. Some CRNs have the property that they deterministically progress to a correct state, no matter the order in which reactions occur. For example, the CRN with the reaction $X \to 2Y$ is guaranteed eventually to reach a state in which the count of Y is twice the initial count of X, i.e., computes the function $f(x) = 2x$, representing the input by species X and the output by species Y. Similarly, the reactions $X_1 \to 2Y$ and $X_2 + Y \to \varnothing$, under arbitrary choice of sequence of the two reactions, compute the function $f(x_1, x_2) = \max\{0, 2x_1 - x_2\}$.

Angluin, Aspnes and Eisenstat [2] investigated the computational behaviour of deterministic CRNs under a different name known as *population protocols* [1]. They showed that the input sets $S \subseteq \mathbb{N}^k$ decidable by deterministic CRNs (i.e. providing "yes" or "no" answers by the presence or absence of certain indicator species) are precisely the *semilinear* subsets of $\mathbb{N}^k$.[1] Chen, Doty, and Soloveichik [5] extended these results to function computation and showed that precisely the semilinear functions (functions f whose graph $\{ (\mathbf{x}, \mathbf{y}) \in \mathbb{N}^{k+l} \mid f(\mathbf{x}) = \mathbf{y} \}$ is a semilinear set) are deterministically computable by CRNs. We say a function $f : \mathbb{N}^k \to \mathbb{N}^l$ is *stably* (a.k.a., *deterministically*) computable by a CRN $\mathcal{C}$ if there are "input" species $X_1, \ldots, X_k$ and "output" species $Y_1, \ldots, Y_l$ such that, if $\mathcal{C}$ starts with $x_1, \ldots, x_k$ copies of $X_1, \ldots, X_k$ respectively, then with probability one, it reaches a count-stable configuration in which the counts of $Y_1, \ldots, Y_l$ are expressed by the vector $f(x_1, ..., x_k)$, and these counts never again change [5].

The method proposed in [5] uses some auxiliary "leader" species present initially, in addition to the input species. To illustrate their utility, suppose that we want to compute function $f(x) = x + 1$ with CRNs. Using the previous approach, we have an input species X (with initial count x), an output species Y and an auxiliary "leader" species L (with initial count 1). The following reactions compute $f(x)$:

$$X \to Y$$
$$L \to Y$$

---

[1] Semilinear sets are defined formally in Section 2. Informally, they are finite unions of "periodic" sets, where the definition of "periodic" is extended in a natural way to multi-dimensional spaces such as $\mathbb{N}^k$.

However, it is experimentally difficult to prepare a solution with a single copy (or a small constant number) of a certain species. The authors of [5] asked whether it is possible to do away with the initial "leader" molecules, i.e., to require that the initial configuration contains initial count $x_1, x_2, \ldots, x_k$ of input species $X_1, X_2, \ldots, X_k$, and initial count 0 of every other species. It is easy to "elect" a single leader molecule from an arbitrary initial number of copies using a reaction such as $L + L \to L$, which eventually reduces the count of L to 1. However, the problem with this approach is that, since L is a reactant in other reactions, there is no way in general to prevent L from participating in these reactions until the reaction $L + L \to L$ has reduced it to a single copy.

Despite these difficulties, we answer the question affirmatively, showing that each semilinear function can be computed by a "leaderless" CRN, i.e., a CRN whose initial configuration contains only the input species. To illustrate one idea used in our construction, consider the function $f(x) = x + 1$ described above. In order to compute the function without a leader (i.e., the initial configuration has x copies of X and 0 copies of every other species), the following reactions suffice:

$$X \to B + 2Y \tag{1.1}$$

$$B + B \to B + K \tag{1.2}$$

$$Y + K \to \varnothing \tag{1.3}$$

Reaction 1.1 produces x copies of B and $2x$ copies of Y. Reaction 1.2 consumes all copies of B except one, so reaction 1.2 executes precisely $x - 1$ times, producing $x - 1$ copies of K. Therefore reaction 1.3 consumes $x - 1$ copies of output species Y, eventually resulting in $2x - (x - 1) = x + 1$ copies of Y. Note that this approach uses a sort of leader election on the B molecules.

In Section 3, we generalize this example, describing a leaderless CRN construction to compute any semilinear function. We use a similar framework to the construction of [5], decomposing the semilinear function into a finite union of affine partial functions (linear functions with an offset; defined formally in Section 2). We show how to compute each affine function with leaderless CRNs, using a fundamentally different construction than the affine-function computing CRNs of [5]. This result, Lemma 3.1, is the primary technical contribution of this paper. Next, in order to decide which affine function should be applied to a given input, we employ the leaderless semilinear predicate computation of Angluin, Aspnes, and Eisenstat [3]; this latter part of the construction is actually identical to the construction of [5], but we include it because our time analysis is different.

Let $n = \|\mathbf{x}\| = \|\mathbf{x}\|_1 = \sum_{i=1}^{k} \mathbf{x}(i)$ be the number of molecules present initially, as well as the volume of the solution. The authors of [5] showed, for each semilinear function f, a direct construction of a CRN that computes f (using leaders) on input $\mathbf{x}$ in expected time $O(n \log n)$. They then combined this direct, error-free construction in parallel with a fast ($O(\log^5 n)$) but error-prone CRN that uses a leader to compute any computable function (including semilinear), using the error-free computation to change the answer of the error-prone computation only if the latter is incorrect. This combination speeds up the computation from expected time $O(n \log n)$ for the direct construction to expected time $O(\log^5 n)$ for the combined construction.

Since we assume no leaders may be supplied in the initial configuration, and since the problem of computing arbitrary computable functions without a leader remains a major open problem [3], this trick does not work for speeding up our construction. However, we show that with some care in the choice of reactions, the direct stable computation of a semilinear function can be done in expected time $O(n)$, improving upon the $O(n \log n)$ bound of the direct construction of [5].

2 Preliminaries

Given a vector $\mathbf{x} \in \mathbb{N}^k$, let $\|\mathbf{x}\| = \|\mathbf{x}\|_1 = \sum_{i=1}^{k} |\mathbf{x}(i)|$, where $\mathbf{x}(i)$ denotes the ith coordinate of $\mathbf{x}$. A set $A \subseteq \mathbb{N}^k$ is *linear* if there exist vectors $\mathbf{b}, \mathbf{u}_1, \ldots, \mathbf{u}_p \in \mathbb{N}^k$ such that

$$A = \{ \, \mathbf{b} + n_1 \mathbf{u}_1 + \ldots + n_p \mathbf{u}_p \mid n_1, \ldots, n_p \in \mathbb{N} \, \}.$$

A is *semilinear* if it is a finite union of linear sets. If $f : \mathbb{N}^k \to \mathbb{N}^l$ is a function, define the *graph* of f to be the set $\{ \, (\mathbf{x}, \mathbf{y}) \in \mathbb{N}^k \times \mathbb{N}^l \mid f(\mathbf{x}) = \mathbf{y} \, \}$. A function is *semilinear* if its graph is semilinear.

We say a partial function $f : \mathbb{N}^k \dashrightarrow \mathbb{N}^l$ is *affine* if there exist kl rational numbers $a_{1,1}, \ldots, a_{k,l} \in \mathbb{Q}$ and $l+k$ nonnegative integers $b_1, \ldots, b_l, c_1, \ldots, c_k \in \mathbb{N}$ such that, if $\mathbf{y} = f(\mathbf{x})$, then for each $j \in \{1, \ldots, l\}$, $\mathbf{y}(j) = b_j + \sum_{i=1}^{k} a_{i,j}(\mathbf{x}(i) - c_i)$, and for each $i \in \{1, \ldots, k\}$, $\mathbf{x}(i) - c_i \geq 0$. In matrix notation, there exist a $k \times l$ rational matrix $\mathbf{A}$ and vectors $\mathbf{b} \in \mathbb{N}^l$ and $\mathbf{c} \in \mathbb{N}^k$ such that $f(\mathbf{x}) = \mathbf{A}(\mathbf{x} - \mathbf{c}) + \mathbf{b}$.

This definition of affine function may appear contrived; see [5] for an explanation of its various intricacies. For reading this paper, the main utility of the definition is that it satisfies Lemma 3.2.

Note that by appropriate integer arithmetic, a partial function $f : \mathbb{N}^k \dashrightarrow \mathbb{N}^l$ is affine if and only if there exist kl integers $n_{1,1}, \ldots, n_{k,l} \in \mathbb{Z}$ and $2l + k$ nonnegative integers $b_1, \ldots, b_l, c_1, \ldots, c_k, d_1, \ldots,$
$d_l \in \mathbb{N}$ such that, if $\mathbf{y} = f(\mathbf{x})$, then for each $j \in \{1, \ldots, l\}$, $\mathbf{y}(j) = b_j + \frac{1}{d_j} \sum_{i=1}^{k} n_{i,j}(\mathbf{x}(i) - c_i)$, and for each $i \in \{1, \ldots, k\}$, $\mathbf{x}(i) - c_i \geq 0$. Each d_j may be taken to be the least common multiple of the denominators of the rational coefficients in the original definition. We employ this latter definition, since it is more convenient for working with integer-valued molecular counts.

2.1 Chemical Reaction Networks

If Λ is a finite set (in this paper, of chemical species), we write $\mathbb{N}^\Lambda$ to denote the set of functions $f : \Lambda \to \mathbb{N}$. Equivalently, we view an element $\mathbf{c} \in \mathbb{N}^\Lambda$ as a vector of $|\Lambda|$ nonnegative integers, with each coordinate "labeled" by an element of Λ. Given $X \in \Lambda$ and $\mathbf{c} \in \mathbb{N}^\Lambda$, we refer to $\mathbf{c}(X)$ as the *count of X in* $\mathbf{c}$.

We write $\mathbf{c} \le \mathbf{c}'$ to denote that $\mathbf{c}(X) \le \mathbf{c}'(X)$ for all $X \in \Lambda$. Given $\mathbf{c}, \mathbf{c}' \in \mathbb{N}^\Lambda$, we define the vector component-wise operations of addition $\mathbf{c} + \mathbf{c}'$, subtraction $\mathbf{c} - \mathbf{c}'$, and scalar multiplication $n\mathbf{c}$ for $n \in \mathbb{N}$. If $\Delta \subset \Lambda$, we view a vector $\mathbf{c} \in \mathbb{N}^\Delta$ equivalently as a vector $\mathbf{c} \in \mathbb{N}^\Lambda$ by assuming $\mathbf{c}(X) = 0$ for all $X \in \Lambda \setminus \Delta$.

Given a finite set of chemical species Λ, a *reaction* over Λ is a triple $\alpha = \langle \mathbf{r}, \mathbf{p}, k \rangle \in \mathbb{N}^\Lambda \times \mathbb{N}^\Lambda \times \mathbb{R}^+$, specifying the stoichiometry of the reactants and products, respectively, and the *rate constant* k. If not specified, assume that $k = 1$ (this is the case for all reactions in this paper), so that the reaction $\alpha = \langle \mathbf{r}, \mathbf{p}, 1 \rangle$ is also represented by the pair $\langle \mathbf{r}, \mathbf{p} \rangle$. For instance, given $\Lambda = \{A, B, C\}$, the reaction $A + 2B \to A + 3C$ is the pair $\langle (1, 2, 0), (1, 0, 3) \rangle$. A *(finite) chemical reaction network (CRN)* is a pair $\mathcal{C} = (\Lambda, R)$, where Λ is a finite set of chemical *species*, and R is a finite set of reactions over Λ. A *configuration* of a CRN $\mathcal{C} = (\Lambda, R)$ is a vector $\mathbf{c} \in \mathbb{N}^\Lambda$. If some current configuration $\mathbf{c}$ is understood from context, we write $\#X$ to denote $\mathbf{c}(X)$.

Given a configuration $\mathbf{c}$ and reaction $\alpha = \langle \mathbf{r}, \mathbf{p} \rangle$, we say that α is *applicable* to $\mathbf{c}$ if $\mathbf{r} \le \mathbf{c}$ (i.e., $\mathbf{c}$ contains enough of each of the reactants for the reaction to occur). If α is applicable to $\mathbf{c}$, then write $\alpha(\mathbf{c})$ to denote the configuration $\mathbf{c} + \mathbf{p} - \mathbf{r}$ (i.e., the configuration that results from applying reaction α to $\mathbf{c}$). If $\mathbf{c}' = \alpha(\mathbf{c})$ for some reaction $\alpha \in R$, we write $\mathbf{c} \to_\mathcal{C} \mathbf{c}'$, or merely $\mathbf{c} \to \mathbf{c}'$ when $\mathcal{C}$ is clear from context. An *execution* (a.k.a., *execution sequence*) $\mathcal{E}$ is a finite or infinite sequence of one or more configurations $\mathcal{E} = (\mathbf{c}_0, \mathbf{c}_1, \mathbf{c}_2, \ldots)$ such that, for all $i \in \{1, \ldots, |\mathcal{E}| - 1\}$, $\mathbf{c}_{i-1} \to \mathbf{c}_i$. If a finite execution sequence starts with $\mathbf{c}$ and ends with $\mathbf{c}'$, we write $\mathbf{c} \to_\mathcal{C}^* \mathbf{c}'$, or merely $\mathbf{c} \to^* \mathbf{c}'$ when the CRN $\mathcal{C}$ is clear from context. In this case, we say that $\mathbf{c}'$ is *reachable* from $\mathbf{c}$.

Turing machines, for example, have different semantic interpretations depending on the computational task under study (deciding a language, computing a function, etc.). Similarly, in this paper we use CRNs to decide subsets of $\mathbb{N}^k$ (for which we reserve the term "chemical reaction *decider*" or CRD) and to compute functions $f : \mathbb{N}^k \to \mathbb{N}^l$ (for which we reserve the term "chemical reaction *computer*" or CRC). In the next two subsections we define two semantic interpretations of CRNs that correspond to these two tasks. We use the term CRN to refer to either a CRD or CRC when the statement is applicable to either type.

These definitions differ slightly from those of [5], because ours are specialized to "leaderless" CRNs: those that can compute a predicate or function in which no species are present in the initial configuration other than the input species. In the terminology of [5], a CRN with species set Λ and input species set Σ is *leaderless* if it has an *initial context* $\sigma : \Lambda \setminus \Sigma \to \mathbb{N}$ such that $\sigma(S) = 0$ for all $S \in \Lambda \setminus \Sigma$. The definitions below are simplified by assuming this to be true of all CRNs.

We also use the convention of Angluin, Aspnes, and Eisenstat [2] that for a CRD, all species "vote" yes or no, rather than only a subset of species as in [5], since this convention is convenient for proving time bounds.

2.2   Stable Decidability of Predicates

We now review the definition of stable decidability of predicates introduced by
Angluin, Aspnes, and Eisenstat [2].[2] Intuitively, the set of species is partitioned
into two sets: those that "vote" yes and those that vote no, and the system stabilizes to an output when a consensus vote is reached (all positive-count species
have the same vote) that can no longer be changed (no species voting the other
way can ever again be produced). It would be too strong to characterize deterministic correctness by requiring all possible executions to achieve the correct
answer; for example, a reversible reaction such as $A \rightleftharpoons B$ could simply be chosen
to run back and forth forever, starving any other reactions. In the more refined
definition that follows, the determinism of the system is captured in that it is
impossible to stabilize to an incorrect answer, and the correct stable output is
always reachable.

A *(leaderless) chemical reaction decider* (CRD) is a tuple $\mathcal{D} = (\Lambda, R, \Sigma, \Upsilon)$,
where (Λ, R) is a CRN, $\Sigma \subseteq \Lambda$ is the *set of input species*, and $\Upsilon \subseteq \Lambda$ is the set of
yes voters, with species in $\Lambda \setminus \Upsilon$ referred to as *no voters*. An input to $\mathcal{D}$ will be an
initial configuration $\mathbf{i} \in \mathbb{N}^{\Sigma}$ (equivalently, $\mathbf{i} \in \mathbb{N}^{k}$ if we write $\Sigma = \{X_1, \ldots, X_k\}$
and assign X_i to represent the i'th coordinate); that is, only input species are
allowed to be non-zero. If we are discussing a CRN understood from context to
have a certain initial configuration $\mathbf{i}$, we write $\#_0 X$ to denote $\mathbf{i}(X)$.

We define a global output partial function $\Phi : \mathbb{N}^{\Lambda} \dashrightarrow \{0, 1\}$ as follows. $\Phi(\mathbf{c})$
is undefined if either $\mathbf{c} = \mathbf{0}$, or if there exist $S_0 \in \Lambda \setminus \Upsilon$ and $S_1 \in \Upsilon$ such that
$\mathbf{c}(S_0) > 0$ and $\mathbf{c}(S_1) > 0$. Otherwise, either $(\forall S \in \Lambda)(\mathbf{c}(S) > 0 \implies S \in \Upsilon)$
or $(\forall S \in \Lambda)(\mathbf{c}(S) > 0 \implies S \in \Lambda \setminus \Upsilon)$; in the former case, the *output* $\Phi(\mathbf{c})$ of
configuration $\mathbf{c}$ is 1, and in the latter case, $\Phi(\mathbf{c}) = 0$.

A configuration $\mathbf{o}$ is *output stable* if $\Phi(\mathbf{o})$ is defined and, for all $\mathbf{c}$ such that
$\mathbf{o} \rightarrow^* \mathbf{c}$, $\Phi(\mathbf{c}) = \Phi(\mathbf{o})$. We say a CRD $\mathcal{D}$ *stably decides* the predicate $\psi : \mathbb{N}^{\Sigma} \rightarrow$
$\{0, 1\}$ if, for any initial configuration $\mathbf{i} \in \mathbb{N}^{k}$, for all configurations $\mathbf{c} \in \mathbb{N}^{\Lambda}$, $\mathbf{i} \rightarrow^* \mathbf{c}$
implies $\mathbf{c} \rightarrow^* \mathbf{o}$ such that $\mathbf{o}$ is output stable and $\Phi(\mathbf{o}) = \psi(\mathbf{i})$. Note that this
condition implies that no incorrect output stable configuration is reachable from
$\mathbf{i}$. We say that $\mathcal{D}$ *stably decides* a set $A \in \mathbb{N}^{k}$ if it stably decides its indicator
function.

The following theorem is due to Angluin, Aspnes, and Eisenstat [2]:

Theorem 2.1 ([2]). *A set $A \subseteq \mathbb{N}^{k}$ is stably decidable by a CRD if and only if
it is semilinear.*

The model they use is defined in a slightly different way; the differences (and
those differences' lack of significance to the questions we explore) are explained
in [5].

---

[2] Those authors use the term "stably *compute*", but we reserve the term "compute" to
apply to the computation of non-Boolean functions. Also, we omit discussion of the
definition of stable computation used in the population protocols literature, which
employs a notion of "fair" executions; the definitions are proven equivalent in [5].

2.3   Stable Computation of Functions

We now define a notion of stable computation of *functions* similar to those above for predicates. Intuitively, the inputs to the function are the initial counts of input species $X_1, \ldots, X_k$, and the outputs are the counts of output species $Y_1, \ldots, Y_l$. The system stabilizes to an output when the counts of the output species can no longer change. Again determinism is captured in that it is impossible to stabilize to an incorrect answer and the correct stable output is always reachable.

A *(leaderless) chemical reaction computer (CRC)* is a tuple $\mathcal{C} = (\Lambda, R, \Sigma, \Gamma)$, where (Λ, R) is a CRN, $\Sigma \subset \Lambda$ is the *set of input species*, $\Gamma \subset \Lambda$ is the *set of output species*, such that $\Sigma \cap \Gamma = \varnothing$. By convention, we let $\Sigma = \{X_1, X_2, \ldots, X_k\}$ and $\Gamma = \{Y_1, Y_2, \ldots, Y_l\}$. We say that a configuration $\mathbf{o}$ is *output stable* if, for every $\mathbf{c}$ such that $\mathbf{o} \to^* \mathbf{c}$ and every $Y_i \in \Gamma$, $\mathbf{o}(Y_i) = \mathbf{c}(Y_i)$ (i.e., the counts of species in Γ will never change if $\mathbf{o}$ is reached). As with CRD's, we require initial configurations $\mathbf{i} \in \mathbb{N}^\Sigma$ in which only input species are allowed to be positive. We say that $\mathcal{C}$ *stably computes* a function $f : \mathbb{N}^k \to \mathbb{N}^l$ if for any initial configuration $\mathbf{i} \in \mathbb{N}^\Sigma$, $\mathbf{i} \to^* \mathbf{c}$ implies $\mathbf{c} \to^* \mathbf{o}$ such that $\mathbf{o}$ is an output stable configuration with $f(\mathbf{i}) = (\mathbf{o}(Y_1), \mathbf{o}(Y_2), \ldots, \mathbf{o}(Y_l))$. Note that this condition implies that no incorrect output stable configuration is reachable from $\mathbf{i}$.

If a CRN stably decides a predicate or stably computes a function, we say the CRN is *stable* (a.k.a., *deterministic*).

2.4   Kinetic Model

The following model of stochastic chemical kinetics is widely used in quantitative biology and other fields dealing with chemical reactions between species present in small counts [9]. It ascribes probabilities to execution sequences, and also defines the time of reactions, allowing us to study the computational complexity of the CRN computation in Section 3.

In this paper, the rate constants of all reactions are 1, and we define the kinetic model with this assumption. The rate constants do not affect the definition of stable computation; they only affect the time analysis. Our time analyses remain asymptotically unaffected if the rate constants are changed (although the constants hidden in the big-O notation would change). A reaction is *unimolecular* if it has one reactant and *bimolecular* if it has two reactants. We use no higher-order reactions in this paper.

The kinetics of a CRN is described by a continuous-time Markov process as follows. Given a fixed volume $v \in \mathbb{R}^+$ and current configuration $\mathbf{c}$, the *propensity* of a unimolecular reaction $\alpha : X \to \ldots$ in configuration $\mathbf{c}$ is $\rho(\mathbf{c}, \alpha) = \mathbf{c}(X)$. The propensity of a bimolecular reaction $\alpha : X + Y \to \ldots$, where $X \neq Y$, is $\rho(\mathbf{c}, \alpha) = \frac{\mathbf{c}(X)\mathbf{c}(Y)}{v}$. The propensity of a bimolecular reaction $\alpha : X + X \to \ldots$ is $\rho(\mathbf{c}, \alpha) = \frac{1}{2}\frac{\mathbf{c}(X)(\mathbf{c}(X)-1)}{v}$. The propensity function determines the evolution of the system as follows. The time until the next reaction occurs is an exponential random variable with rate $\rho(\mathbf{c}) = \sum_{\alpha \in R} \rho(\mathbf{c}, \alpha)$ (note that $\rho(\mathbf{c}) = 0$ if no reactions are applicable to $\mathbf{c}$). Therefore, the expected time for the next reaction to occur is $\frac{1}{\rho(\mathbf{c})}$.

The kinetic model is based on the physical assumption of well-mixedness valid in a dilute solution. Thus, we assume the *finite density constraint*, which stipulates that a volume required to execute a CRN must be proportional to the maximum molecular count obtained during execution [14]. In other words, the total concentration (molecular count per volume) is bounded. This realistically constrains the speed of the computation achievable by CRNs. Note, however, that it is problematic to define the kinetic model for CRNs in which the reachable configuration space is unbounded for some start configurations, because this means that arbitrarily large molecular counts are reachable.[3] We apply the kinetic model only to CRNs with configuration spaces that are bounded for each start configuration, choosing the volume to be equal to the reachable configuration with the highest molecular count (in this paper, this will always be within a constant multiplicative factor of the number of input molecules).

It is not difficult to show that if a CRN is stable and has a finite reachable configuration space from any initial configuration $\mathbf{i}$, then under the kinetic model (in fact, for any choice of rate constants), with probability 1 the CRN will eventually reach an output stable configuration.

We require the following lemmas, whose proofs we omit in this extended abstract.

Lemma 2.2. Let $\mathcal{A} = \{A_1, \ldots, A_m\}$ be a set of species with the property that they appear only in applicable reactions of the form $A_i \rightarrow \sum_l B_l$, where $B_l \notin \mathcal{A}$. Then starting from a configuration $\mathbf{c}$ in which for all $i \in \{1, \ldots, m\}$, $\mathbf{c}(A_i) = O(n)$, with volume $O(n)$, the expected time to reach a configuration in which none of the described reactions can occur is $O(\log n)$.

Lemma 2.3. Let $\mathcal{A} = \{A_1, \ldots, A_m\}$ be a set of species with the property that they appear only in applicable reactions of the form $A_i + A_j \rightarrow A_k + \sum_l B_l$, where $B_l \notin \mathcal{A}$, and for all $i, j \in \{1, \ldots, m\}$, there is at least one reaction $A_i + A_j \rightarrow \ldots$. Then starting from a configuration $\mathbf{c}$ in which for all $i \in \{1, \ldots, m\}$, $\mathbf{c}(A_i) = O(n)$, with volume $O(n)$, the expected time to reach a configuration in which none of the described reactions can occur is $O(n)$.

Lemma 2.4. Let $\mathcal{C} = \{C_1, \ldots, C_m\}$ and $\mathcal{A} = \{A_1, \ldots, A_p\}$ be two sets of species with the property that they appear only in applicable reactions of the form $C_i + A_j \rightarrow C_i + \sum_l B_l$, where $B_l \notin \mathcal{A}$. Then starting from a configuration $\mathbf{c}$ in which for all $i \in \{1, \ldots, m\}$, $\mathbf{c}(C_i) = \Omega(n)$, and for all $j \in \{1, \ldots, p\}$, $\mathbf{c}(A_j) = O(n)$, with volume $O(n)$, the expected time to reach a configuration in which none of the described reactions can occur is $O(\log n)$.

3 Leaderless CRCs Can Compute Semilinear Functions

To supply an input vector $\mathbf{x} \in \mathbb{N}^k$ to a CRN, we use an initial configuration with $\mathbf{x}(i)$ molecules of input species X_i. Throughout this section, we let $n = ||\mathbf{x}||_1 =$

---

[3] One possibility is to have a "dynamically" growing volume as in [14].

$\sum_{i=1}^{k} \mathbf{x}(i)$ denote the initial number of molecules in solution. Since all CRNs we employ have the property that they produce at most a constant multiplicative factor more molecules than are initially present, this implies that the volume required to satisfy the finite density constraint is $O(n)$.

Suppose the CRC $\mathcal{C}$ stably computes a function $f : \mathbb{N}^k \dashrightarrow \mathbb{N}^l$. We say that $\mathcal{C}$ stably computes f *monotonically* if its output species are not consumed in any reaction.[4]

We show in Lemma 3.1 that affine partial functions can be computed in expected time $O(n)$ by a leaderless CRC. For its use in proving Theorem 3.4, we require that the output molecules be produced monotonically. If we used a direct encoding of the output of the function, this would be impossible for general affine functions. For example, consider the function $f(x_1, x_2) = x_1 - x_2$ where $\text{dom } f = \{ (x_1, x_2) \mid x_1 \geq x_2 \}$. By withholding a single copy of X_2 and letting the CRC stabilize to the output value $\#Y = x_1 - x_2 + 1$, then allowing the extra copy of X_2 to interact, the only way to stabilize to the correct output value $x_1 - x_2$ is to consume a copy of the output species Y. Therefore Lemma 3.1 computes f indirectly via an encoding of f's output that allows monotonic production of outputs, encoding the output value $\mathbf{y}(j)$ as the difference between the counts of two monotonically produced species Y_j^P and Y_j^C, a concept formalized by the following definition.

Let $f : \mathbb{N}^k \dashrightarrow \mathbb{N}^l$ be a partial function. We say that a partial function $\hat{f} : \mathbb{N}^k \dashrightarrow \mathbb{N}^l \times \mathbb{N}^l$ is a *diff-representation* of f if $\text{dom } f = \text{dom } \hat{f}$ and, for all $\mathbf{x} \in \text{dom } f$, if $(\mathbf{y}_P, \mathbf{y}_C) = \hat{f}(\mathbf{x})$, where $\mathbf{y}_P, \mathbf{y}_C \in \mathbb{N}^l$, then $f(\mathbf{x}) = \mathbf{y}_P - \mathbf{y}_C$, and $\mathbf{y}_P = O(f(\mathbf{x}))$. In other words, $\hat{f}$ represents f as the difference of its two outputs $\mathbf{y}_P$ and $\mathbf{y}_C$, with the larger output $\mathbf{y}_P$ possibly being larger than the original function's output, but at most a multiplicative constant larger.

The following lemma is the main technical result required for proving our main theorem, Theorem 3.4. It shows that every affine function can be computed (via a diff-representation) in time $O(n)$ by a leaderless CRC.

Lemma 3.1. *Let $f : \mathbb{N}^k \dashrightarrow \mathbb{N}^l$ be an affine partial function. Then there is a diff-representation $\hat{f} : \mathbb{N}^k \dashrightarrow \mathbb{N}^l \times \mathbb{N}^l$ of f and a leaderless CRC that monotonically stably computes $\hat{f}$ in expected time $O(n)$.*

Proof. If f is affine, then there exist kl integers $n_{1,1}, \ldots, n_{k,l} \in \mathbb{Z}$ and $2l + k$ nonnegative integers $b_1, \ldots, b_l, c_1, \ldots, c_k, d_1, \ldots, d_l \in \mathbb{N}$ such that, if $\mathbf{y} = f(\mathbf{x})$, then for each $j \in \{1, \ldots, l\}$, $\mathbf{y}(j) = b_j + \frac{1}{d_j} \sum_{i=1}^{k} n_{i,j}(\mathbf{x}(i) - c_i)$, and for each $i \in \{1, \ldots, k\}$, $\mathbf{x}(i) - c_i \geq 0$. Define the CRC as follows. It has input species $\Sigma = \{X_1, \ldots, X_k\}$ and output species $\Gamma = \{Y_1^P, \ldots, Y_l^P, Y_1^C, \ldots, Y_l^C\}$.

There are three main components of the CRN, separately handling the c_i offset, the $n_{i,j}/d_j$ coefficient, and the b_j offset.

The latter two components both make use of Y_j^C molecules to account for production of Y_j^P molecules in excess of $\mathbf{y}(j)$ to ensure that $\#_\infty Y_j^P - \#_\infty Y_j^C =$

---

[4] Its output species could potentially be reactants so long as they are catalytic, meaning that the stoichiometry of the species as a product is at least as great as its stoichiometry as a reactant, e.g. $X + Y \to Z + Y$ or $A + Y \to Y + Y$.

$\mathbf{y}(j)$, which establishes that the CRC stably computes a diff-representation of f. It is clear by inspection of the reactions that $\#_\infty Y_j^P = O(\mathbf{y}(j))$.

Add the reaction

$$X_1 \to C_{1,1} + B_1 + B_2 + \ldots + B_l + b_1 Y_1^P + b_2 Y_2^P + \ldots b_l Y_l^P \qquad (3.1)$$

The first product $C_{1,1}$ will be used to handle the c_1 offset, and the remaining products will be used to handle the b_j offsets. For each $i \in \{2, \ldots, k\}$, add the reaction

$$X_i \to C_{i,1} \qquad (3.2)$$

By Lemma 2.2, reactions (3.1) and (3.2) take time $O(\log n)$ to complete.

We now describe the three components of the CRC separately.

$\underline{c_i \text{ offset:}}$ Reactions (3.1) and (3.2) produce $\mathbf{x}(i)$ copies of $C_{i,1}$. We must reduce this number by c_i, producing $\mathbf{x}(i) - c_i$ copies of X_i', the species that will be used by the next component to handle the $n_{i,j}/d_j$ coefficient. A high-order reaction implementing this is $(c_i + 1)C_{i,1} \to c_i C_{i,1} + X_i'$, since that reaction will eventually happen exactly $\mathbf{x}(i) - c_i$ times (stopping when $\#C_{i,1}$ reaches c_i). This is implemented by the following bimolecular reactions. For each $i \in \{1, \ldots, k\}$ and $m, p \in \{1, \ldots, c_i\}$, if $m + p \le c_i$, add the reaction

$$C_{i,m} + C_{i,p} \to C_{i,m+p}.$$

If $m + p > c_i$, add the reaction

$$C_{i,m} + C_{i,p} \to C_{i,c_i} + (m + p - c_i)X_i'.$$

By Lemma 2.3, these reactions complete in expected time $O(n)$.

$\underline{n_{i,j}/d_j \text{ coefficient:}}$ For each $i \in \{1, \ldots, k\}$, add the reaction

$$X_i' \to X_{i,1} + X_{i,2} + \ldots + X_{i,l}$$

This allows each output to be associated with its own copy of the input. By Lemma 2.2, these reactions complete in expected time $O(\log n)$.

For each $i \in \{1, \ldots, k\}$ and $j \in \{1, \ldots, l\}$, add the reaction

$$X_{i,j} \to \begin{cases} n_{i,j} D_{j,1}^P, & \text{if } n_{i,j} > 0; \\ (-n_{i,j}) D_{j,1}^C, & \text{if } n_{i,j} < 0. \end{cases}$$

By Lemma 2.2, these reactions complete in expected time $O(\log n)$.

We must now divide $\#D_{j,1}^P$ and $\#D_{j,1}^C$ by d_j. This is accomplished by the high-order reactions $d_j D_{j,1}^P \to Y_j^P$ and $d_j D_{j,1}^C \to Y_j^C$. Similarly to the previous component, we implement these with the following reactions for $d_j \ge 1$. We first handle the case $d_j > 1$. For each $j \in \{1, \ldots, l\}$ and $m, p \in \{1, \ldots, d_j - 1\}$, if $m + p \le d_j - 1$, add the reactions

$$D_{j,m}^P + D_{j,p}^P \to D_{j,m+p}^P$$
$$D_{j,m}^C + D_{j,p}^C \to D_{j,m+p}^C$$

If $m + p > c_i$, add the reactions

$$D_{j,m}^P + D_{j,p}^P \to D_{j,m+p-d_j}^P + Y_j^P$$
$$D_{j,m}^C + D_{j,p}^C \to D_{j,m+p-d_j}^C + Y_j^C$$

By Lemma 2.3, these reactions complete in expected time $O(n)$.
When $d_j = 1$, we only have the following unimolecular reactions.

$$D_{j,1}^P \to Y_j^P$$
$$D_{j,1}^C \to Y_j^C$$

By Lemma 2.2, these reactions complete in expected time $O(\log n)$.
These reactions will produce $\frac{1}{d_j} \sum_{n_{i,j}>0} n_{i,j}(\mathbf{x}(i) - c_i)$ copies of Y_j^P and $-\frac{1}{d_j} \sum_{n_{i,j}<0} n_{i,j}(\mathbf{x}(i) - c_i)$ copies of Y_j^C. Therefore, letting $\#_{\mathrm{coef}} Y_j^P$ and $\#_{\mathrm{coef}} Y_j^C$ denote the number of copies of Y_j^P and Y_j^C eventually produced just by this component, it holds that $\#_{\mathrm{coef}} Y_j^P - \#_{\mathrm{coef}} Y_j^C = \frac{1}{d_j} \sum_{i=1}^k n_{i,j}(\mathbf{x}(i) - c_i)$.

b_j **offset:** For each $j \in \{1, \ldots, l\}$, add the reaction

$$B_j + B_j \to B_j + b_j Y_j^C \tag{3.3}$$

By Lemma 2.3, these reactions complete in expected time $O(n)$.
Reaction (3.1) produces b_j copies of Y_j^P for each copy of B_j produced, which is $\mathbf{x}(i)$. Reaction (3.3) occurs precisely $\mathbf{x}(i) - 1$ times. Therefore reaction (3.3) produces precisely b_j fewer copies of Y_j^C than reaction (3.1) produces of Y_j^P. This implies that when all copies of Y_j^C are eventually produced by reaction (3.3), the number of Y_j^P's produced by reaction (3.1) minus the number of Y_j^C's produced by reaction (3.3) is b_j.   □

We require the following lemma, proven in [5].

Lemma 3.2 ([5]). *Let $f : \mathbb{N}^k \to \mathbb{N}^l$ be a semilinear function. Then there is a finite set $\{f_1 : \mathbb{N}^k \dashrightarrow \mathbb{N}^l, \ldots, f_m : \mathbb{N}^k \dashrightarrow \mathbb{N}^l\}$ of affine partial functions, where each $\mathrm{dom}\, f_i$ is a linear set, such that, for each $\mathbf{x} \in \mathbb{N}^k$, if $f_i(\mathbf{x})$ is defined, then $f(\mathbf{x}) = f_i(\mathbf{x})$, and $\bigcup_{i=1}^m \mathrm{dom}\, f_i = \mathbb{N}^k$.*

We require the following theorem, due to Angluin, Aspnes, and Eisenstat [3, Theorem 5], which states that any semilinear predicate can be decided by a CRD in expected time $O(n)$.

Theorem 3.3 ([3]). *Let $\phi : \mathbb{N}^k \to \{0, 1\}$ be a semilinear predicate. Then there is a leaderless CRD $\mathcal{D}$ that stably decides ϕ, and the expected time to reach an output-stable configuration is $O(n)$.*

The following is the main theorem of this paper. It shows that semilinear functions can be computed by leaderless CRCs in linear expected time.

Theorem 3.4. *Let* $f : \mathbb{N}^k \to \mathbb{N}^l$ *be a semilinear function. Then there is a leaderless CRC that stably computes* f *in expected time* $O(n)$.

Proof. The CRC will have input species $\Sigma = \{X_1, \ldots, X_k\}$ and output species $\Gamma = \{Y_1, \ldots, Y_l\}$. By Lemma 3.2, there is a finite set $F = \{f_1 : \mathbb{N}^k \dashrightarrow \mathbb{N}^l, \ldots, f_m : \mathbb{N}^k \dashrightarrow \mathbb{N}^l\}$ of affine partial functions, where each dom f_i is a linear set, such that, for each $\mathbf{x} \in \mathbb{N}^k$, if $f_i(\mathbf{x})$ is defined, then $f(\mathbf{x}) = f_i(\mathbf{x})$. We compute f on input $\mathbf{x}$ as follows. Since each dom f_i is a linear (and therefore semilinear) set, by Theorem 3.3 we compute each semilinear predicate $\phi_i = $ "$\mathbf{x} \in$ dom f_i and $(\forall i' \in \{1, \ldots, i-1\})$ $\mathbf{x} \notin$ dom $f_{i'}$?" by separate parallel CRD's each stabilizing in expected time $O(n)$. (The latter condition ensures that for each $\mathbf{x}$, precisely one of the predicates is true, in case the domains of the partial functions have nonempty intersection.)

By Lemma 3.1, for each $i \in \{1, \ldots, m\}$, there is a diff-representation $\hat{f}_i$ of f_i that can be stably computed by parallel CRC's. Assume that for each $i \in \{1, \ldots, m\}$ and each $j \in \{1, \ldots, l\}$, the jth pair of outputs $\mathbf{y}_P(j)$ and $\mathbf{y}_C(j)$ of the ith function is represented by species $\hat{Y}_{i,j}^P$ and $\hat{Y}_{i,j}^C$. We interpret each $\hat{Y}_{i,j}^P$ and $\hat{Y}_{i,j}^C$ as an "inactive" version of "active" output species $Y_{i,j}^P$ and $Y_{i,j}^C$.

For each $i \in \{1, \ldots, m\}$, for the CRD $\mathcal{D}_i = (\Lambda, R, \Sigma, \Upsilon)$ computing the predicate ϕ_i, let L_i^1 represent any species in Υ, and L_i^0 represent any species in $\Lambda \setminus \Upsilon$, and that once $\mathcal{D}_i$ reaches an output stable configuration, $\#L_i^b = \Omega(n)$, where b is the output of $\mathcal{D}_i$. Then add the following reactions for each $i \in \{1, \ldots, m\}$ and each $j \in \{1, \ldots, l\}$:

$$L_i^1 + \hat{Y}_{i,j}^P \to L_i^1 + Y_{i,j}^P + Y_j \tag{3.4}$$

$$L_i^0 + Y_{i,j}^P \to L_i^0 + M_{i,j} \tag{3.5}$$

$$M_{i,j} + Y_j \to \hat{Y}_{i,j}^P \tag{3.6}$$

The latter two reactions implement the reverse direction of the first reaction – using L_i^0 as a catalyst instead of L_i^1 – using only bimolecular reactions. Also add the reactions

$$L_i^1 + \hat{Y}_{i,j}^C \to L_i^1 + Y_{i,j}^C \tag{3.7}$$

$$L_i^0 + Y_{i,j}^C \to L_i^0 + \hat{Y}_{i,j}^C \tag{3.8}$$

and

$$Y_{i,j}^P + Y_{i,j}^C \to K_j \tag{3.9}$$

$$K_j + Y_j \to \varnothing \tag{3.10}$$

That is, a "yes" answer for function i activates the ith output and a "no" answer deactivates the ith output. Eventually each CRD stabilizes so that precisely one i has L_i^1 present, and for all $i' \neq i$, $L_{i'}^0$ is present. We now claim that at this point, all outputs for the correct function $\hat{f}_i$ will be activated and all other outputs will be deactivated. The reactions enforce that at any time, $\#Y_j = \#K_j + \sum_{i=1}^{m}(\#Y_{i,j}^P + \#M_{i,j})$. In particular, $\#Y_j \geq \#K_j$ and $\#Y_j \geq \#M_{i,j}$ at all times, so there will never be a K_j or $M_{i,j}$ molecule that cannot participate in the reaction of which it is a reactant. Eventually $\#Y_{i,j}^P$ and $\#Y_{i,j}^C$ stabilize to 0 for all but one value of i (by reactions (3.5), (3.6), (3.8)), and for this value of i, $\#Y_{i,j}^P$ stabilizes to $\mathbf{y}(j)$ and $\#Y_{i,j}^C$ stabilizes to 0 (by reaction (3.9)). Eventually $\#K_j$ stabilizes to 0 by the last reaction. Eventually $\#M_{i,j}$ stabilizes to 0 since L_i^0 is absent for the correct function $\hat{f}_i$. This ensures that $\#Y_j$ stabilizes to $\mathbf{y}(j)$.

It remains to analyze the expected time to stabilization. Let $n = \|\mathbf{x}\|$. By Lemma 3.1, the expected time for each affine function computation to complete is $O(n)$. Since the $\hat{Y}_{i,j}^P$ are produced monotonically, the most $Y_{i,j}^P$ molecules that are ever produced is $\#_\infty \hat{Y}_{i,j}^P$. Since we have m computations in parallel, the expected time for all of them to complete is $O(nm) = O(n)$ (since m depends on f but not n). We must also wait for each predicate computation to complete. By Theorem 3.3, each of these predicates takes expected time $O(n)$ to complete, so all of them complete in expected time $O(mn) = O(n)$.

At this point, the L_1^i leaders must convert inactive output species to active, and $L_0^{i'}$ (for $i' \neq i$) must convert active output species to inactive. By Lemma 2.4, reactions (3.4), (3.5), (3.7), and (3.8) complete in expected time $O(\log n)$. Once this is completed, by Lemma 2.3, reaction (3.6) completes in expected time $O(n)$. Reaction (3.9) completes in expected time $O(n)$ by Lemma 2.3. Once this is done, reaction (3.10) completes in expected time $O(n)$ by Lemma 2.3. $\square$

4   Conclusion

The clearest shortcoming of our leaderless CRC, compared to the leader-employing CRC of [5], is the time complexity. Our CRC takes expected time $O(n)$ to complete with n input molecules, versus $O(\log^5 n)$ for the CRC of Theorem 4.4 of [5]. However, we do obtain a modest speedup ($O(n)$ versus $O(n\log n)$), compared to the *direct* construction of Theorem 4.1 of [5]. The indirect construction of Theorem 4.1 of [5] relied heavily on the use of a fast, error-prone CRN which computes arbitrary computable functions, and which crucially uses a leader. The major open question is, for each semilinear function $f : \mathbb{N}^k \to \mathbb{N}^l$, is there a leaderless CRC that stably computes f on input of size n in expected time $t(n)$, where t is a sublinear function? This may relate to the question of whether there is a sublinear time CRN that solves the leader election problem, i.e., in volume n with an initial state with n copies of species X and no other species initially present, produce a single copy of a species L. However, it is conceivable that there is a direct way to compute semilinear functions quickly without needing to use a leader election.

If this is not possible for all semilinear functions, another interesting open question is to precisely characterize the class of functions that can be stably computed by a leaderless CRC in polylogarithmic time. For example, the class of linear functions with positive integer coefficients (e.g., $f(x_1, x_2) = 3x_1 + 2x_2$) has this property since they are computable by $O(\log n)$-time unimolecular reactions such as $X_1 \to 3Y, X_2 \to 2Y$. However, most of the CRN programming techniques used to generalize beyond such functions seem to require some bimolecular reaction $A + B \to \ldots$ in which it is possible to have $\#A = \#B = 1$, making the expected time at least n just for this reaction.

Acknowledgement. We are indebted to Anne Condon for helpful discussions and suggestions.

References

1. Angluin, D., Aspnes, J., Diamadi, Z., Fischer, M., Peralta, R.: Computation in networks of passively mobile finite-state sensors. Distributed Computing 18, 235–253 (2006); Preliminary version appeared in PODC 2004
2. Angluin, D., Aspnes, J., Eisenstat, D.: Stably computable predicates are semilinear. In: PODC 2006: Proceedings of the Twenty-fifth Annual ACM Symposium on Principles of Distributed Computing, pp. 292–299. ACM Press, New York (2006)
3. Angluin, D., Aspnes, J., Eisenstat, D.: Fast computation by population protocols with a leader. In: Dolev, S. (ed.) DISC 2006. LNCS, vol. 4167, pp. 61–75. Springer, Heidelberg (2006)
4. Cardelli, L.: Strand algebras for DNA computing. Natural Computing 10(1), 407–428 (2011)
5. Chen, H.-L., Doty, D., Soloveichik, D.: Deterministic function computation with chemical reaction networks. In: Stefanovic, D., Turberfield, A. (eds.) DNA 18. LNCS, vol. 7433, pp. 25–42. Springer, Heidelberg (2012)
6. Condon, A., Hu, A., Maňuch, J., Thachuk, C.: Less haste, less waste: On recycling and its limits in strand displacement systems. Journal of the Royal Society Interface 2, 512–521 (2012); In: Cardelli, L., Shih, W. (eds.) DNA 17. LNCS, vol. 6937, pp. 84–99. Springer, Heidelberg (2011)
7. Condon, A., Kirkpatrick, B., Maňuch, J.: Reachability bounds for chemical reaction networks and strand displacement systems. In: Stefanovic, D., Turberfield, A. (eds.) DNA 18. LNCS, vol. 7433, pp. 43–57. Springer, Heidelberg (2012)
8. Cook, M., Soloveichik, D., Winfree, E., Bruck, J.: Programmability of chemical reaction networks. In: Condon, A., Harel, D., Kok, J.N., Salomaa, A., Winfree, E. (eds.) Algorithmic Bioprocesses, pp. 543–584. Springer, Heidelberg (2009)
9. Gillespie, D.T.: Exact stochastic simulation of coupled chemical reactions. Journal of Physical Chemistry 81(25), 2340–2361 (1977)
10. Hjelmfelt, A., Weinberger, E.D., Ross, J.: Chemical implementation of neural networks and Turing machines. Proceedings of the National Academy of Sciences 88(24), 10983–10987 (1991)
11. Jiang, H., Riedel, M., Parhi, K.: Digital signal processing with molecular reactions. IEEE Design and Test of Computers 29(3), 21–31 (2012)
12. Magnasco, M.O.: Chemical kinetics is Turing universal. Physical Review Letters 78(6), 1190–1193 (1997)

13. Soloveichik, D.: Robust stochastic chemical reaction networks and bounded tau-leaping. Journal of Computational Biology 16(3), 501–522 (2009)
14. Soloveichik, D., Cook, M., Winfree, E., Bruck, J.: Computation with finite stochastic chemical reaction networks. Natural Computing 7(4), 615–633 (2008)
15. Thachuk, C., Condon, A.: Space and energy efficient computation with DNA strand displacement systems. In: Stefanovic, D., Turberfield, A. (eds.) DNA 18. LNCS, vol. 7433, pp. 135–149. Springer, Heidelberg (2012)
16. Zavattaro, G., Cardelli, L.: Termination problems in chemical kinetics. In: van Breugel, F., Chechik, M. (eds.) CONCUR 2008. LNCS, vol. 5201, pp. 477–491. Springer, Heidelberg (2008)

DNA Sticky End Design and Assignment for Robust Algorithmic Self-assembly

Constantine G. Evans[1] and Erik Winfree[2]

[1] Physics,
[2] Computer Science,
California Institute of Technology

Abstract. A major challenge in practical DNA tile self-assembly is the minimization of errors. Using the kinetic Tile Assembly Model, a theoretical model of self-assembly, it has been shown that errors can be reduced through abstract tile set design. In this paper, we instead investigate the effects of "sticky end" sequence choices in systems using the kinetic model along with the nearest-neighbor model of DNA interactions. We show that both the sticky end sequences present in a system and their positions in the system can significantly affect error rates, and propose algorithms for sequence design and assignment.

1 Introduction

Self-assembly of DNA tiles is a promising technique for the assembly of complex nanoscale structures. Assembly of tiles can be programmed by designing short complementary single-stranded DNA "sticky ends." While assembly using unique tile types or simple lattices is often studied [26,16], algorithmic growth, where small sets with few tile types can form complex assemblies, is particularly powerful theoretically, and has been studied extensively through the abstract Tile Assembly Model (aTAM) [28,8,17].

A number of different designs for tile structure are used for assembly [26,21,16]. As an example, the DAO-E tile design (Fig. 1(a)) consists of two helices connected by two crossovers, with four 5 nucleotide (nt) sticky ends, one at each end of each helix. Experimentally, conditions are usually used such that tiles will favorably attach by two bonds between sticky-end regions, adding cooperativity to binding. In the abstract Tile Assembly Model, this is modelled by individual tiles attaching to edges of the current assembly when they can make at least two correct bonds to adjacent tiles ($T = 2$), and never detaching once attached.

The Pascal mod 3 (PM3) system shown in Fig. 1(b) is a simple example. The tiles implement addition modulo 3, akin to Pascal's triangle. Tiles attach by their two lower-left ends, and then provide ends for future tiles to attach that sum the logical values of the two "input" ends. Growth proceeds to the upper-right, controlled by a V-shaped seed of tiles that attach by strength-2 bonds and provide edges of logical 1s.

A more sophisticated example, the counter system from Barish et al [3], is shown in Fig. 1(c). In this system, a ribbon of tiles grows from a large seed

D. Soloveichik and B. Yurke (Eds.): DNA 2013, LNCS 8141, pp. 61–75, 2013.

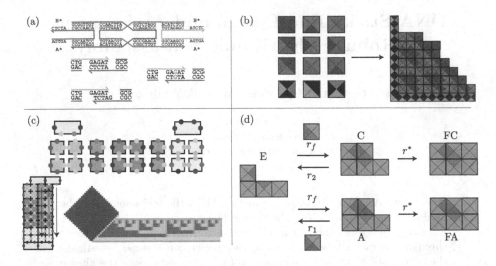

Fig. 1. Tile systems, structures and the kinetic trapping model. (a) shows an example DAO-E tile structure [21], along with examples of complementary and partially mismatched sticky end attachments. (b) shows the Pascal mod 3 tile system along with a potential perfect assembly. Blue, green and red correspond to ends with logical values 0, 1, and 2, respectively, while black indicates double-strength bonds of the V-shaped seed. (c) shows the tiles (top) in the Barish counter system, along with an illustration of zig-zag ribbon growth (left) and an Xgrow simulation of growth from an origami seed (blue), where each pixel represents one tile. Orange and brown tiles indicate tiles with logical values of 1 and 0, respectively, while gray tiles are boundary and nucleation barrier tiles, and incrementing tiles are green. (d) illustrates the states and transition rates in the kinetic trapping model of growth errors.

structure of DNA origami. Rows of tiles grow in a zig-zag fashion, with each new row being started by a double tile that is equivalent to two permanently-attached single tiles. On "downward" rows tiles increment a bit string with two tiles per bit from the previous row, while on "upward" rows, corresponding tiles copy the newly-incremented row. These tiles implement a binary counter starting from whatever bit string was specified on the original origami seed and incrementing every two rows of tiles.

In examining algorithmic growth of experimental systems, the kinetic Tile Assembly Model provides better physical relevance [28]. Tiles are assumed to be in solution at a particular concentration, which is usually assumed to be constant. Tiles attach to empty lattice sites at a rate r_f dependent only on their concentration, and detach at a rate r_b ($b = 1, 2, \ldots$) dependent upon the number of correct "sticky end" attachments they have to the assembly:

$$r_f = \hat{k}e^{-G_{mc}} \qquad\qquad r_b = \hat{k}e^{-bG_{se}} . \qquad (1)$$

Here G_{mc} is a dimensionless free energy analogue related to tile concentration by $[c] = e^{-G_{mc}+\alpha}$, G_{se} is the sign-reversed dimensionless free energy of a single bond, b is the number of correct bonds, and $\hat{k}$ is an adjusted forward rate constant

$\hat{k} \equiv k_f e^{\alpha}$, where k_f is the usual second-order mass action rate constant for tile attachment, typically $k_f = 10^6$ /M/s. This model has been used for numerous theoretical and computational simulation studies of algorithmic tile assembly [29,6,10,8,17], and has fit well with experimental findings both qualitatively and quantitatively [9,11].

As the kinetic model allows any tile to attach regardless of correctness, it is challenging to design tile systems that exhibit algorithmic behavior while keeping erroneous growth low enough to obtain high yields of correct assemblies. Growth errors in the kinetic model are well studied, and often modelled by the kinetic trapping model. The model considers tiles attaching and detaching at a single lattice location, while having a rate for an attached tile to become "frozen" in place by further growth. This rate, $r^* = \hat{k}e^{-G_{mc}} - \hat{k}e^{-2G_{se}}$, is related to the overall growth rate of the system [28]. As tiles that attach without any correct bonds ("doubly-mismatched" tiles) will detach very quickly, to first approximation, the only states that need to be considered are empty (E), correct tile (C), and "almost correct tile" (A)—a tile that is attached by one correct bond—along with frozen states for correct and almost correct tiles (FC and FA). These states are described in Fig. 1(d).

Numerous techniques have been studied to reduce such error rates, especially "proofreading" transformations that transform individual tiles into multiple tile blocks or sets of tiles [29,6,4,20]. These techniques have been shown to significantly reduce error rates both in simulation and experimentally [11,6,3]. Such techniques rely on changing tile systems at an abstract level, and reduce error rates of even ideal systems. However, in implementing the abstract logic of a tile system in actual DNA tiles, design complexities cause the system's kinetics to deviate from the default kTAM parameters. In particular, the single-stranded "sticky ends" that implement the abstract ends must be chosen from a finite sequence space to be both as uniform in binding energy and as orthogonal as possible. Deviations here can introduce further errors [10].

2   Theoretical Model

In the kTAM, G_{se} and G_{mc} are by default considered to be constant and independent of both tiles and sticky ends. A more detailed model cannot assume this. G_{mc} is dependent upon tile concentration: the value may be different for each tile type, and may change as free tiles are depleted by attachment. However, as experimental techniques exist to keep tile concentrations approximately constant throughout assembly [23], we will assume a time-invariant (but possibly tile type dependent) G_{mc}.

G_{se}, on the other hand, will depend upon the bonds between sticky ends. Ends with different sequences will have different free energies for binding to their complements, and some ends may be able to partially bind to ends that are only partially complementary (Fig. 1(a)). This results in a G_{se}^{ij} for each pair of sticky ends (i, j). In the default kTAM, all non-diagonal terms will be zero, and all diagonal terms will be equal. G_{se}^{ij} can thus be defined in terms of deviations from a reference G_{se}:

$$G_{se}^{ii} = G_{se} + \delta_i \qquad\qquad G_{se}^{ij} = s_{ij}G_{se} \text{ for } i \neq j . \qquad (2)$$

Non-uniform sticky ends, with non-zero δ_i, will affect the detachment rate of correct and almost-correct tile attachments, while spurious non-orthogonal binding strengths s_{ij} will only decrease detachment rates for almost-correct and doubly-mismatched tile attachments. In the following theoretical analysis, the much lower likelihood doubly-mismatched interactions are ignored. For simulations, done with the Xgrow kTAM simulator [2], these interactions are taken into account when there is non-orthogonal binding.

2.1   Uniformity

Non-uniform sticky end energies have been simulated previously [10], but have not been studied analytically. In the kTAM, the growth rate of an assembly depends on the difference between on and off rates [28], which we approximate for a uniform system as $r^* = \hat{k}e^{-G_{mc}} - \hat{k}e^{-2G_{se}}$.

For a system with non-uniform energies, a tile attaching by two i bonds will have

$$r^* = \hat{k}e^{-G_{mc}} - \hat{k}e^{-2G_{se}-2\delta_i} = \hat{k}e^{-2G_{se}} \left(e^\epsilon - e^{-2\delta_i}\right)$$

where we define $\epsilon \equiv 2G_{se} - G_{mc}$, a measure of supersaturation: for an ideal system, $\epsilon = 0$ results in unbiased growth, whereas $\epsilon > 0$ results in forward growth and $\epsilon < 0$ causes crystals to shrink. As can be seen, the growth rate will depend on the δ_i's of the bonds in the growth region. With $\delta_i < -\frac{1}{2}\epsilon$ (negative δ corresponds to weaker binding), growth in a region won't be favorable.

In the worst case, where tiles attaching by two bonds with the smallest δ_i form a sufficiently large region, growth can only be ensured if $\epsilon > -2 \min\{\delta_i\}$, and error rates can be approximated by the kTAM with this minimum ϵ value. The kinetic trapping model in the default kTAM results in an error rate $P_{error} \approx me^{-G_{se}+\epsilon}$ for m possible incorrect tile attachments [28], so the worst-case error rate for a given $\delta_{\min} \equiv \min\{\delta_i\}$ would be

$$P_{\text{error}} \approx me^{-G_{se}-2\delta_{\min}} . \qquad (3)$$

Fig. 2(a) shows simulations of the PM3 system with ϵ adjusted along the lines of our worst-case growth requirements. For positive deviations, where most ends remain at the same strength, assembly time is largely unchanged, while the error rate increases. For negative (weaker bond) deviations, where ϵ is adjusted, the error rate rises per Eq. 3, while the assembly time decreases sharply as most tiles attach with the same G_{se}^{ii} but are at a higher concentration.

While this method to adjust tile concentrations ensures crystal growth, it may not obtain the optimal trade-off between growth rate and error rate. This trade-off has been addressed for perfect sticky ends [5,12], but is more complicated with imperfect sticky ends and complex tile sets. Rather than simply adjusting all concentrations uniformly, the assumption can be made, which is not necessarily optimal, that error rates for a complex tile set can be reduced by ensuring

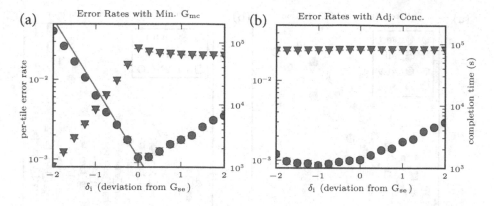

Fig. 2. Error rates for Pascal mod 3 systems with non-uniform end interactions simulated in Xgrow. In both (a) and (b), single sticky ends have been changed so that $G_{se}^{ii} = G_{se} + \delta_i$, while all others have remained at G_{se}. In (a), the ϵ for the system has been uniformly changed to always allow forward-growth by two of the weakest bond types by setting G_{mc}. In (b), the tiles with deviating ends have had their concentration adjusted so that all tiles have the same growth rate $r^* = \hat{k}e^{-G_{mc}^n} - \hat{k}e^{-G_{se}^{ii} - G_{se}^{jj}}$, where tile type n attaches using sticky end types i and j. Blue circles show error rates; green triangles show the time taken to construct an 8000 tile assembly, the line in (a) shows Eq. 3. For these simulations, we set base parameters of $G_{se} = 10$ and $G_{mc} = 19.2$.

that the overall growth rate remains uniform throughout the crystal. This can be achieved by modifying the concentrations of tiles to modify their G_{mc} values such that the r^* for each tile type is the same. Fig. 2(b) shows simulations of this form of concentration-adjustment with the PM3 system. As expected, assembly time remains almost completely unchanged across a large range of deviations. Meanwhile negative deviations do not significantly increase error rates, and positive deviations increase error rates in a manner similar to Fig. 2(a).

2.2 Orthogonality

Unlike non-uniformity, the kinetic trapping model for growth errors can be easily extended to account for non-orthogonality. Assuming $s_{ij} \ll 1$, growth errors will be primarily caused by almost-correct tiles attaching by one correct and one incorrect bond, as in the ideal case. A uniform incorrect bond strength of s, and m possible almost-correct tiles for a given lattice site, then gives the following rates of change between the different states shown in Fig. 1(d):

$$
\dot{P}(t) = \begin{array}{c} E \\ C \\ A \\ FC \\ FA \end{array}
\begin{pmatrix}
-2r_f & r_2 & r_{(1+s)} & 0 & 0 \\
r_f & -r_2 - r^* & 0 & 0 & 0 \\
mr_f & 0 & -r_{(1+s)} - r^* & 0 & 0 \\
0 & r^* & 0 & 0 & 0 \\
0 & 0 & r^* & 0 & 0
\end{pmatrix} P(t) \; .
\tag{4}
$$

Here $P(t)$ is a vector of probabilities at time t that the site will be in a state $[E, C, A, FC, FA]$. The steady state of this is not useful, as any combination of

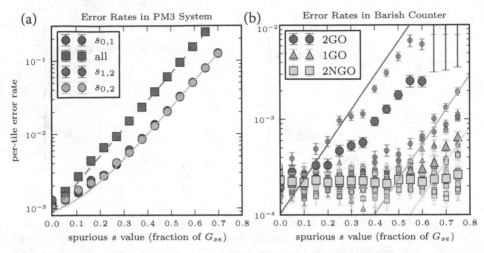

Fig. 3. Error rates with non-orthogonal interactions. (a) shows interactions for the PM3 system; circles and solid lines show simulated and theoretical error rates, respectively, with single pairs interacting. Squares and dashed lines show error rates for a uniform non-orthogonal interaction between every pair. (b) shows error rates for sensitive single non-orthogonal pairs in the Barish counter system, along with lines showing $e^{-(s-\sigma)G_{se}}$ for various values of σ chosen to roughly follow the worst pairs of each sensitivity. Small dots represent individual pairs, while large dots show averages for sensitivity classes. For (a) $G_{se} = 10$ and $G_{mc} = 19.2$, for (b) $G_{se} = 8.35$ and $G_{mc} = 17.8$.

FC and FA will be a steady state. Instead, the eventual probability of being in FA after starting only in state E at $t = 0$ will provide an error rate per additional tile in an assembly. This can be treated as a flow problem, where we consider the differential accumulation into FC and FA from E, as in Winfree [28]. From this, the probability of an almost-correct tile being trapped in place is:

$$P_{error} = \frac{m}{m + \frac{r_f + r_{1+s}}{r_f + r_2}} \approx \frac{1}{1 + \frac{1}{m}e^{(1-s)G_{se}-\epsilon}} \approx m e^{(s-1)G_{se}+\epsilon} \ . \tag{5}$$

While tile systems will have a different number of possible almost-correct tiles for different lattice sites, making this result less applicable, the PM3 system has an equal number for every possible lattice site. Fig. 3(a) shows error rates in simulations with interactions between single pairs of ends and for a uniform non-orthogonal interaction energy between every pair. In both cases, error rates largely follow Eq. 5.

2.3   Sticky End Sensitivity

When non-orthogonal sticky end interactions are not uniform, the degree of their influence on error rates may depend on which tile types they appear on and the logical interactions within the tile set. In systems where a tile never has the opportunity to attach with strength $1 + s_{ij}$, interactions between i and j may be less relevant, whereas other pairs of ends in the system may allow tiles to erroneously attach during correct growth and be simply locked in place by continued growth. For example, Fig. 3(b) shows error rates for the Barish counter system when non-orthogonal interactions are introduced between single pairs of sticky ends. These pairs have been organized into sets (1NGO, 2NGO, 1GO, and 2GO) based on a model described below of how interactions between them may affect the tile system. As can be seen, this model has some success in predicting the impact different pairs will have on error rates.

We start by assuming that all attachments in growth occur with single tiles attaching by exactly two correct strength-1 bonds. Assuming that each tile in the system can have its ends labelled as inputs or outputs, and that every growth site has a unique tile that can attach by inputs, all lattice locations possible in the system will eventually be filled by a specific tile. Rather than looking at lattice sites that actually appear in correct growth, which would require simulation, we can combinatorially investigate all possible local neighborhoods that might appear, and conservatively examine them for possible problems. For example, whether there exists a tile that can attach with strength $1 + s_{ij}$ can be approximated by whether there are two tiles that share a common input bond on one side but not the other, so that when one tile incorrectly attaches where the other could attach correctly, it forms a strength 1 bond for the common bond and a strength s_{ij} bond for the mismatch (as in Fig. 4(a)).

We describe end pairs where such tiles exist as being in the set of "first-order sensitive" end pairs. If the sides of the tiles are inputs for at least one tile type, and thus the tiles can attach in normal forward growth, the end pair is in the set of first-order growth oriented sensitive (1GO) pairs, whereas without consideration of input and output sides, the end pair is in the set of first-order non-growth-oriented sensitive (1NGO) pairs. End pairs (i, j) that are in 1NGO but not 1GO have tiles that can attach with strength $1 + s_{ij}$ only during growth after an error or at sites where there is no correct tile.

While end pairs in these sets have tiles that allow the first, erroneous tile attachment in the kinetic trapping model, the model also requires that a second tile be able attach by two correct bonds to the erroneous tile and adjacent tiles to trap the error in place. This is also not necessarily possible: an incorrect attachment could result in there being no adjacent correct attachment, and designing systems where this is the case is in fact the goal of proofreading systems [29].

Thus we can devise "second-order sensitive" sets of end pairs that allow this second, correct tile attachment, and are therefore expected to be more likely to cause errors. Consider a pair of tiles A and X with a common bond on one side but not the other, satisfying the criteria for a first-order sensitive pair. Whether a further tile can attach with strength 2 can be approximated by whether there is some second pair of tiles, B and Y, that can each attach to some third side of their respective original tiles, and also share a common bond on another side. In a plausible local neighborhood where A and B could attach correctly in sequence, it is possible for X to first attach erroneously, with strength $1 + s_{ij}$ (in the location

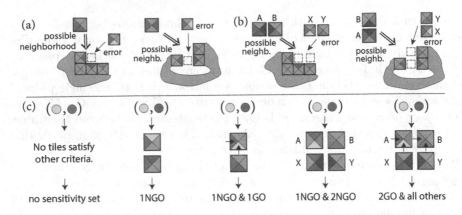

Fig. 4. Illustration of end pair sensitivity sets. For simplicity, all left and bottom sides are considered inputs. (a) shows, for given tiles, examples of possible local neighborhoods they could attach to and tiles that could erroneously attach via first-order sensitivity. (b) shows, for given pairs of tiles A and B, examples of local neighborhoods the pair could attach to in sequence, and a pair of tiles X and Y that could erroneously attach via second-order sensitivity. (c) shows examples of tiles satisfying various criteria for the shown end pairs to be in different sensitivity sets; arrows show examples of required input sides for growth-oriented sets.

where A could have bound), then for Y to attach with strength 2 (where B could have bound after A) owing to the second commond bond, as in Fig. 4(b).

As with first-order sensitivity, if the common and differing sides of the first pair of tiles are inputs, and sides of the second pair of tiles that are shared or attach to the first pair are also inputs, then the end pair involved is in the set of second-order growth oriented sensitive (2GO) pairs, whereas without consideration of inputs, the pair is in the set of second-order non-growth-oriented sensitive (2NGO) pairs.

These sets can be summarized more formally as follows, while examples of satisfying tiles are shown in Fig. 4(c):

- An end pair (i, j) is in the set of first-order sensitive end pairs if there exist at least two tiles in the tile system where both tiles share a common end k on one side, and on some other side, one tile has end i and the other has end j. If at least one of the two tiles has k and either i or j as inputs, then the end pair is in 1GO and 1NGO, otherwise, it is only in 1NGO.
- To determine if a first-order sensitive end pair (i, j) is in the set of second-order sensitive end pairs, consider a pair of tiles that satisfy the first-order criteria, and additional pairs of tiles that can attach to the first pair by bonds l and m (possibly the same) on a third side. If there exist a pair of these additional tiles that also share a common bond n, then the end pair is second-order sensitive. If at least one of the first tiles has k and either i or j as inputs, and one of the additional tiles attaching to it has n and either l or m as an input, then the end pair is in 2GO and 2NGO, otherwise, it is only in 2NGO.

Note that this analysis is done without determining what assemblies and thus what local neighborhoods actually form, so the combinations of inputs being considered might never appear during the growth of a correct assembly. As such, it is conceivable that, for example, an end pair could be in 2GO without ever having an effect in correct growth of an assembly. While this is a significant limitation, determining if a combination of inputs ever occurs, or if two tiles are ever assembled adjacent to each other, is in general undecidable by reduction to the Halting problem [27]. Furthermore, our current software treats all bonds as strength-1, and all tiles as single tiles, with double tiles being represented by a pair of single tiles with a fake bond that is then excluded from the sets; whilst the set definitions could be extended to account for double tiles and strength-2 bonds, we have not yet investigated the complexities involved.

Also, while pairs may be in either or both of 1GO or 2NGO, in all systems we have considered, all pairs in 1GO have also been in 2NGO, and there have been no pairs that are only in 1NGO. End pairs that aren't in *any* of these sets, and can be described as "zeroth-order," should have interactions between them that have a negligible effect on error rates in the kinetic trapping model.

Very rough theoretical estimates of the contributions that sensitive end pairs will have on a system can be obtained by considering the number of tiles that need to attach incorrectly. For pairs in 2GO, as only the initial tile will need to attach incorrectly before it can be locked in place by a correct attachment, the probablity of an error every time such a situation occurs is $\sim e^{(s-1)Gse}$. For those in 1GO but not 2GO, since there is no correct attachment after the first tile attaches incorrectly, at least one further incorrect attachment will be required, giving a probability of error $\sim e^{(s-2)Gse}$ or lower. For pairs only in 2NGO or 1NGO, the probability that the first tile can attach incorrectly will depend upon the likelihood that growth is proceeding in an incorrect direction, which in turn will depend upon numerous factors, but will usually require at least one previous incorrect attachment, giving another factor of $\sim e^{-Gse}$ on top of their GO counterparts.

For the Barish counter, there are 342 pairs of ends (helix direction prevents around half the ends from attaching to the other half). Of these, 22 are 2NGO, 9 are both 1GO and 2NGO, and 3 are also 2GO. Fig. 3(b) shows error rates for increasing values of s_{ij} where one pair has its value increased and all other spurious pairs are left with $s_{ij} = 0$. Each pair has been classified by its "worst" set. As can be seen, 2NGO pairs have little impact on error rates beyond those seen in the ideal kTAM, 1GO pairs start to have an effect after around $s_{ij} > 0.4$, and 2GO pairs are the most sensitive. In the case of the three 2GO pairs in the Barish counter, two cause errors that prevent correct growth in the next row without an additional error, explaining the significant difference between the most sensitive 2GO pair and the two less sensitive pairs.

3   Sequence Design and Assignment

3.1   Sequence Design

DNA sequence set design for molecular computation is a widely-studied problem. Different applications necessitate different constraints and approaches: longer sequences with less stringent requirements can be constrained with combinatorial methods like Hamming distance [13], while work on sequences with more stringent requirements have used thermodynamic constraints [25]. However, the basic goal shared throughout most of these algorithms is to find the largest set of DNA sequences that hybridize to their complements significantly better than to any other sequences in the set, or to find a set of a certain size with the best possible "quality"; in this the problem is similar to the maximum independent set problem, which is NP complete [7,18].

For sticky ends, the sequence lengths required, especially the 5 to 6 nt ends of DAO-E tiles, are shorter, and provide a smaller sequence space, than most other work has considered, with a few exceptions that have largely generated very small sets [25]. Using the end pair sensitivity model, we can reduce errors from non-orthogonal interactions by changing the assignment of sequences to abstract ends, as described later. However, we have no corresponding model to allow us to compensate for non-uniform energies.

The goal for our sequence design, therefore, is to find a requested number of sequences that (a) have non-orthogonal interactions less than a set constraint, and (b) have binding energies (melting temperatures) as uniform as possible given the orthogonality constraints. This contrasts with many sequence design algorithms, where a minimum melting temperature is of primary importance [24], and from algorithms that simply constrain melting temperatures to be within set constraints [25], in that our algorithm chooses a sequence with the *closest* melting temperature at each step.

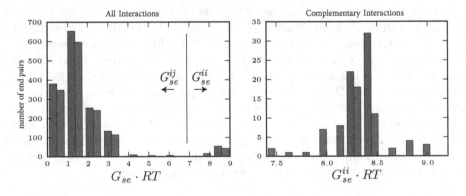

Fig. 5. Histograms of end pair interactions with the original Barish sequences (red) and newly designed sequences (blue). (a) shows all end pairs, (b) shows a zoomed-in area containing all end-complement pairs. All energies were calculated using the energy model in our sequence designer at 37 °C.

As the lengths of sticky end sequences are short, complex secondary structure is limited, and thus our algorithm uses an approximation of minimum free energy (MFE) for thermodynamic calculations. Similar to the "h-measure" used in Phan et al [18], the algorithm considers hybridization between two sequences with every possible offset, and uses the nearest-neighbor interaction data from SantaLucia et al [22], including values for symmetric loops, dangles, single-base mismatched pairs, and coaxial stacking with core sequences. Furthermore, for DAO-E tiles, core helix bases adjacent to the sticky ends affect energetics, and need to be designed alongside the sticky end sequences.

Our algorithm works as follows, for length L sticky ends.

1. Generate a set of all possible available sequences A that fit user requirements. With adjacent bases considered, this could be as many as 4^{L+2} sequences.
2. Calculate end-complement binding energies $G_{se}^{ii'}$ for all sequences in A, and (to speed up computation) remove any sequence that falls outside a user-specified range around the median $G_{se}^{ii'}$ of all sequences initially in A, which we call $\overline{G_{se}}$.
3. For each sequence needed:
   (a) Randomly choose a sequence i from all sequences in A that are closest to $\overline{G_{se}}$, and add this to the set of chosen sequences C.
   (b) Calculate the G_{se}^{ij} between i and every remaining sequence j in A, and remove all sequences from A with a G_{se}^{ij} greater than a user-specified value.
4. Stop when either A is empty, or a sufficient number of sequences have been generated.

$\overline{G_{se}}$ is chosen as the desired ideal G_{se} in order to ensure a large number of sequences with similar G_{se}^{ii}s will be available, for 5 nt ends, the desired value is $G_{se} \cdot RT = 8.35$ kcal/mol at $37°C$. By adjusting parameters, the maximum number of sequences that can be chosen can be changed as shown in Table 1; running the algorithm repeatedly will also find different numbers of sequences.

Sets chosen by this algorithm are guaranteed to have all ends interact less than a set amount $s_{ij} < s_{\text{desired}}$ with ends other than their complements, and to deviate from the desired correct interaction by less than a set amount $|\delta_i| < \delta_{\text{desired}}$, though when generating sets of a fixed size the largest δ_is will often be much smaller, as the software selects for the smallest δ_i values possible.

Fig. 5 shows a comparison between end pair interactions in the original Barish counter system and new sequences designed with our sequence design software. As can be seen, our software prevents large non-orthogonal interactions of 4 kcal/mol $< G_{se}^{ij} \cdot RT < 6$ kcal/mol, but does not significantly reduce interactions with $G_{se}^{ij} \cdot RT < 4$ kcal/mol. However, for complementary interactions, our software is able to find a significantly more uniform set of ends.

The practical value of this designer depends on the accuracy of the underlying energy model, of course, but the same algorithm can be used with different energy models as understanding of sticky end energetics is improved. The algorithm, with some energy model modifications, may also be of use in other

Table 1. Examples of the number of sticky ends found by our designer for varying user-specified parameters (bold). For lengths 5 and 6, examples are the best out of 100 runs, while for length 10, the example is a single run.

Length (nt)	$\overline{G_{se}} \cdot RT$	$\max(s_{ij})$	# found	$\mathrm{std}(\delta_i)$	$\max \delta_i$
5	8.354	0.2	5	$0.04G_{se}$	$0.1G_{se}$
5	8.354	0.4	21	$0.01G_{se}$	$0.038G_{se}$
5	8.354	0.5	40	$0.01G_{se}$	$0.036G_{se}$
6	9.818	0.4	29	$0.004G_{se}$	$0.015G_{se}$
10	15.454	0.4	183	$0.01G_{se}$	$0.05G_{se}$

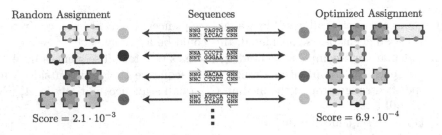

Fig. 6. Illustration of end assignment for the Barish counter set with new sequences. For conciseness, only a portion of the ends are shown.

areas of DNA computation where very short sequences with very similar melting temperatures and low non-orthogonal interactions are needed, such as toehold regions in strand displacement systems. However, it does not consider a number of factors important for actual strand displacement regions, and starts to become computationally intractable for sequences longer than 10 or 11 nt.

3.2 Sequence Assignment

The sequence designer is able to find sets of ends with very similar complementary interactions, and low non-orthogonal interactions. However, by ensuring that sequences are assigned to ends in a system such that end pairs with higher sensitivity have lower interactions, errors can further be reduced, and perhaps more importantly, the chance that a poor choice of sequences is made for a critical pair of ends can be minimized.

We assigned ends using a simulated annealing algorithm that used, as a score, the sum of rough error estimates for each end pair (see Fig. 4):

$$
S(\text{assignment}) = \sum_{i,j \in 2GO} e^{-(s_{ij}-1.1)G_{se}} + \sum_{i,j \in 1GO \text{ and } \notin 2GO} e^{-(s_{ij}-1.5)G_{se}} \tag{6}
$$
$$
+ \sum_{i,j \in 2NGO \text{ and } \notin 1GO} e^{-(s_{ij}-1.65)G_{se}} + \sum_{i,j \in 1NGO \text{ and } \notin 2NGO} e^{-(s_{ij}-2)G_{se}} .
$$

We call the resulting assignment 'optimized', although of course it is not guaranteed to be a global optimum. Offset values in the exponents were set

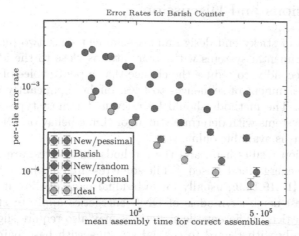

Fig. 7. Error rates for the Barish counter system with different sticky end sequences. Error rates are calculated from the percentage of correct assemblies formed of size 673. G_{se} values are calculated from ends, or are uniformly $G_{se} \cdot RT = 8.35$ kcal/mol in the ideal case. G_{mc} values were varied between 17.6 and 17.9. 1000 simulations were run for each G_{mc} value.

by rough estimates of the worst errors for different classes in the simulations shown in Fig. 4, and terms here for 2GO, 1GO and 2NGO are shown by solid lines in that figure. For 1NGO, the -2 parameter is chosen simply to be lower than other classes, as no system we have examined has end pairs that are only 1NGO. Since the sequence designer chooses adjacent bases as well as sticky end sequences, sequences can be consistently assigned to ends on all tiles, as in Fig. 6. The sequences and tiles for the Barish counter cannot be assigned in the same way, as different tiles with the same sticky end types often have different adjacent base pairs, modifying their interactions. Furthermore, as the sequence assignment algorithm only considers non-orthogonal interactions, results on a system with significant non-uniformity will likely be inconsistent.

Fig. 7 shows simulated error rates and assembly time for counters using sequences from Barish et al [3], sequences designed by our sequence designer and randomly assigned, and the same designed sequences assigned by our simulated annealing algorithm to both minimize and maximize the score in Eq. 6, along with error rates and assembly time for the system under ideal kTAM conditions. For a range of G_{mc} values and resultant assembly times, there is at least a 3-fold improvement in error rate between new sequences that are pessimally and optimally assigned by our scoring function, with increasing improvement as the assembly rate, and thus ideal error rate, decreases. For optimally assigned sequences, error remains close to the ideal error rate. The original sequences and assignment for the Barish counter perform slightly better than the pessimally assigned new sequences.

4   Conclusions and Discussion

These methods of sticky end design and assignment serve two purposes: firstly, to design experimental systems with error rates as close to the ideal kTAM as possible, and secondly, to reduce the chance that a poor choice of sequences, or even a poor assignment of sequences to tiles, might significantly impact experimental results. The methods should be relevant for most types of DNA tiles, and most tile systems with deterministic algorithmic behavior. Our software for these algorithms is available online [1].

The simulation results here, and the methods themselves, are reliant on the accuracy of the energy model used. While some research has been done on sticky-end energetics [15,19,14,9], usually for individual pairs of tiles, it is not known how well nearest-neighbor models of DNA energetics apply to sticky ends on DNA tiles in lattices. Different tile structures may also require slightly different models, especially with regard to coaxial stacking with base pairs adjacent to the sticky ends.

It is possible that extending end sensitivity definitions to higher orders, considering more than two tile attachments, may be a useful area of investigation, especially when considering tile systems making use of similarly higher order proofreading. Indeed, proofreading can counteract at a more fundamental level some of the same errors that arise from non-orthogonal interactions. The effects of non-uniform sticky end energies, however, may still significantly impact proofreading sets, and remain a potentially fruitful area of research beyond our simplistic modeling and concentration adjustment technique.

References

1. StickyDesign, http://dna.caltech.edu/StickyDesign/
2. The Xgrow simulator, http://dna.caltech.edu/Xgrow/
3. Barish, R.D., Schulman, R., Rothemund, P.W.K., Winfree, E.: An information-bearing seed for nucleating algorithmic self-assembly. Proc. Natl. Acad. Sci. USA 106, 6054–6059 (2009)
4. Chen, H.-L., Goel, A.: Error free self-assembly using error prone tiles. In: Ferretti, C., Mauri, G., Zandron, C. (eds.) DNA10. LNCS, vol. 3384, pp. 62–75. Springer, Heidelberg (2005)
5. Chen, H.-L., Kao, M.-Y.: Optimizing tile concentrations to minimize errors and time for DNA tile self-assembly systems. In: Sakakibara, Y., Mi, Y. (eds.) DNA16. LNCS, vol. 6518, pp. 13–24. Springer, Heidelberg (2011)
6. Chen, H.-L., Schulman, R., Goel, A., Winfree, E.: Reducing facet nucleation during algorithmic self-assembly. Nano Lett. 7, 2913–2919 (2007)
7. Deaton, R., Chen, J., Bi, H., Rose, J.: A software tool for generating non-crosshybridizing libraries of DNA oligonucleotides. In: Hagiya, M., Ohuchi, A. (eds.) DNA8. LNCS, vol. 2568, pp. 252–261. Springer, Heidelberg (2003)
8. Doty, D.: Theory of algorithmic self-assembly. Commun. ACM 55, 78–88 (2012)
9. Evans, C.G., Hariadi, R.F., Winfree, E.: Direct atomic force microscopy observation of DNA tile crystal growth at the single-molecule level. J. Am. Chem. Soc. 134, 10485–10492 (2012)

10. Fujibayashi, K., Murata, S.: Precise simulation model for DNA tile self-assembly. IEEE Trans. Nanotechnol. 8, 361–368 (2009)
11. Fujibayashi, K., Hariadi, R.F., Park, S.H., Winfree, E., Murata, S.: Toward reliable algorithmic self-assembly of DNA tiles: A fixed-width cellular automaton pattern. Nano Lett. 8, 1791–1797 (2008)
12. Jang, B., Kim, Y., Lombardi, F.: Error tolerance of DNA self-assembly by monomer concentration control. In: 2006 21st IEEE International Symposium on Defect and Fault Tolerance in VLSI Systems, pp. 89–97. IEEE (2006)
13. Kick, A., Bönsch, M., Mertig, M.: EGNAS: An exhaustive DNA sequence design algorithm. BMC Bioinformatics 13, 138 (2012)
14. Li, Z., Liu, M., Wang, L., Nangreave, J., Yan, H., Liu, Y.: Molecular behavior of DNA origami in higher-order self-assembly. J. Am. Chem. Soc. 132, 13545–13552 (2013)
15. Nangreave, J., Yan, H., Liu, Y.: Studies of thermal stability of multivalent DNA hybridization in a nanostructured system. Biophys. J. 97, 563–571 (2009)
16. Park, S.H., Yin, P., Liu, Y., Reif, J.H., LaBean, T.H., Yan, H.: Programmable DNA self-assemblies for nanoscale organization of ligands and proteins. Nano Lett. 5, 729–733 (2013)
17. Patitz, M.J.: An introduction to tile-based self-assembly. In: Durand-Lose, J., Jonoska, N. (eds.) UCNC 2012. LNCS, vol. 7445, pp. 34–62. Springer, Heidelberg (2012)
18. Phan, V., Garzon, M.H.: On codeword design in metric DNA spaces. Nat. Comput. 8, 571–588 (2008)
19. Pinheiro, A.V., Nangreave, J., Jiang, S., Yan, H., Liu, Y.: Steric crowding and the kinetics of DNA hybridization within a DNA nanostructure system. ACS Nano 6, 5521–5530 (2013)
20. Reif, J.H., Sahu, S., Yin, P.: Compact error-resilient computational DNA tiling assemblies. In: Ferretti, C., Mauri, G., Zandron, C. (eds.) DNA10. LNCS, vol. 3384, pp. 293–307. Springer, Heidelberg (2005)
21. Rothemund, P.W.K., Papadakis, N., Winfree, E.: Algorithmic self-assembly of DNA Sierpinski triangles. PLoS Biol. 2, e424 (2004)
22. SantaLucia, J., Hicks, D.: The Thermodynamics of DNA Structural Motifs. Annu. Rev. Biophys. Biomol. Struct. 33, 415–440 (2004)
23. Schulman, R., Yurke, B., Winfree, E.: Robust self-replication of combinatorial information via crystal growth and scission. Proc. Natl. Acad. Sci. USA 109, 6405–6410 (2012)
24. Tanaka, F.: Design of nucleic acid sequences for DNA computing based on a thermodynamic approach. Nucleic Acids Res. 33, 903–911 (2005)
25. Tulpan, D., Andronescu, M., Chang, S.B., Shortreed, M.R., Condon, A., Hoos, H.H., Smith, L.M.: Thermodynamically based DNA strand design. Nucleic Acids Res. 33, 4951–4964 (2005)
26. Wei, B., Dai, M., Yin, P.: Complex shapes self-assembled from single-stranded DNA tiles. Nature 485, 623–626 (2012)
27. Winfree, E.: On the Computational Power of DNA Annealing and Ligation. In: DNA Computers. DIMACS Series in Discrete Mathematics and Computer Science, pp. 199–221. AMS (1996)
28. Winfree, E.: Simulations of computing by self-assembly. Tech. Rep. Caltech CSTR:1998.22, California Inst. Technol., Pasadena, CA (1998)
29. Winfree, E., Bekbolatov, R.: Proofreading tile sets: Error correction for algorithmic self-assembly. In: Chen, J., Reif, J.H. (eds.) DNA9. LNCS, vol. 2943, pp. 126–144. Springer, Heidelberg (2004)

DNA Reservoir Computing:
A Novel Molecular Computing Approach

Alireza Goudarzi[1], Matthew R. Lakin[1], and Darko Stefanovic[1,2]

[1] Department of Computer Science
University of New Mexico
[2] Center for Biomedical Engineering
University of New Mexico
alirezag@cs.unm.edu

Abstract. We propose a novel molecular computing approach based on reservoir computing. In reservoir computing, a dynamical core, called a *reservoir*, is perturbed with an external input signal while a *readout layer* maps the reservoir dynamics to a target output. Computation takes place as a transformation from the input space to a high-dimensional spatiotemporal feature space created by the transient dynamics of the reservoir. The readout layer then combines these features to produce the target output. We show that coupled deoxyribozyme oscillators can act as the reservoir. We show that despite using only three coupled oscillators, a molecular reservoir computer could achieve 90% accuracy on a benchmark temporal problem.

1   Introduction

A reservoir computer is a device that uses transient dynamics of a system in a critical regime—a regime in which perturbations to the system's trajectory in its phase space neither spread nor die out—to transform an input signal into a desired output [1]. We propose a novel technique for molecular computing based on the dynamics of molecular reactions in a microfluidic setting. The dynamical core of the system that contains the molecular reaction is called a reservoir. We design a simple *in-silico* reservoir computer using a network of deoxyribozyme oscillators [2], and use it to solve temporal tasks. The advantage of this method is that it does not require any specific structure for the reservoir implementation except for rich dynamics. This makes the method an attractive approach to be used with emerging computing architectures [3].

We choose deoxyribozyme oscillators due to the simplicity of the corresponding mathematical model and the rich dynamics that it produces. In principle, the design is generalizable to any set of reactions that show rich dynamics. We reduce the oscillator model in [2] to a form more amenable to mathematical analysis. Using the reduced model, we show that the oscillator dynamics can be easily tuned to our needs. The model describes the oscillatory dynamics of three product and three substrate species in a network of three coupled oscillators. We introduce the input to the oscillator network by fluctuating the supply of substrate molecules and we train a *readout layer* to map the oscillator dynamics onto a target output. For a complete physical reservoir computing design, two main problems should be addressed: (1) physical implementation of the

D. Soloveichik and B. Yurke (Eds.): DNA 2013, LNCS 8141, pp. 76–89, 2013.

reservoir and (2) physical implementation of the readout layer. In this paper, we focus on a chemical design for the reservoir and assume that the oscillator dynamics can be read using fluorescent probes and processed using software. We aim to design a complete chemical implementation of the reservoir and the readout layer in a future work (cf. Section 5). A similar path was taken by Smerieri et al. [4] to achieve an all-analog reservoir computing design using an optical reservoir introduced by Paquot et al. [5].

We use the molecular reservoir computer to solve two temporal tasks of different levels of difficulty. For both tasks, the readout layer must compute a function of past inputs to the reservoir. For Task A, the output is a function of two immediate past inputs, and for Task B, the output is a function of two past inputs, one τ seconds ago and the other $\frac{3}{2}\tau$ seconds ago. We implement two varieties of reservoir computer, one in which the readout layer only reads the dynamics of product concentrations and another in which both product and substrate concentrations are read. We show that the product-only version achieves about 70% accuracy on Task A and about 80% accuracy on Task B, whereas the product-and-substrate version achieves about 80% accuracy on Task A and 90% accuracy on Task B. The higher performance on Task B is due to the longer time delay, which gives the reservoir enough time to process the input. Compared with other reservoir computer implementations, the molecular reservoir computer performance is surprisingly good despite the reservoir being made of only three coupled oscillators.

2 Reservoir Computing

As reservoir computing (RC) is a relatively new paradigm, we try to convey the sense of how it computes and explain why it is suitable for molecular computing. RC achieves computation using the dynamics of an excitable medium, the reservoir [6]. We perturb the intrinsic dynamics of the reservoir using a time-varying input and then read and translate the traces of the perturbation on the system's trajectory onto a target output.

RC was developed independently by Maass et al. [7] as a model of information processing in cortical microcircuits, and by Jaeger [8] as an alternative approach to time-series analysis using Recurrent Neural Networks (RNN). In the RNN architecture, the nodes are fully interconnected and learning is achieved by updating all the connection weights [8, 9]. However, this process is computationally very intensive. Unlike the regular structure in RNN, the reservoir in RC is built using sparsely interconnected nodes, initialized with fixed random weights. There are input and output layers which feed the network with inputs and obtain the output, respectively. To get the desired output, we have to compute only the weights on the connections from the reservoir to the output layer using examples of input-output sequence.

Figure 1 shows a sample RC architecture with sparse connectivity between the input and the reservoir, and between the nodes inside the reservoir. The output node is connected to all the reservoir nodes. The input weight matrix is an $I \times N$ matrix $\mathbf{W}^{in} = [w_{i,j}^{in}]$, where I is the number of input nodes, N is the number of nodes in the reservoir, and $w_{j,i}^{in}$ is the weight of the connection from input node i to reservoir node j. The connection weights inside the reservoir are represented by an $N \times N$ matrix $\mathbf{W}^{res} = [w_{j,k}^{res}]$, where

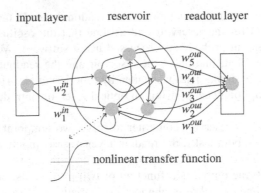

input layer   reservoir   readout layer

nonlinear transfer function

Fig. 1. Schematic of a generic reservoir computer. The input is weighted and then fed into a reservoir made up of a number of nodes with nonlinear transfer functions. The nodes are interconnected using the coupling matrix $\mathbf{W}^{res} = [w_{ij}^{res}]$, where w_{ij}^{res} is the weight from node j to node i. The weights are selected randomly from identical and independent distributions. The output is generated using linear combination of the values of the nodes in the reservoir using output weight vector $\mathbf{W}^{out} = [w_i^{out}]$.

$w_{j,k}^{res}$ is the weight from node k to node j in the reservoir. The output weight matrix is an $N \times O$ matrix $\mathbf{W}^{out} = [w_{l,k}^{out}]$, where O is the number of output nodes and $w_{l,k}^{out}$ is the weight of the connection from reservoir node k to output node l. All the weights are samples of i.i.d. random variables, usually taken to be normally distributed with mean $\mu = 0$ and standard deviation σ. We can tune μ and σ depending on the properties of $U(t)$ to achieve optimal performance. We represent the time-varying input signal by an Ith order column vector $\mathbf{U}(t) = [u_i(t)]$, the reservoir state by an Nth order column vector $\mathbf{X}(t) = [x_j(t)]$, and the generated output by an Oth order column vector $\mathbf{Y}(t) = [y_l(t)]$. We compute the time evolution of each reservoir node in discrete time as:

$$x_j(t+1) = f(\mathbf{W}_j^{res} \cdot \mathbf{X}(t) + \mathbf{W}^{in} \cdot \mathbf{U}(t)), \tag{1}$$

where f is the nonlinear transfer function of the reservoir nodes, $\cdot$ is the matrix dot product, and $\mathbf{W}_j^{res}$ is the jth row of the reservoir weight matrix. The reservoir output is then given by:

$$\mathbf{Y}(t) = w_b + \mathbf{W}^{out} \cdot \mathbf{X}(t), \tag{2}$$

where w_b is an inductive bias. One can use any regression method to train the output weights to minimize the output error $E = ||\mathbf{Y}(t) - \widehat{\mathbf{Y}}(t)||^2$ given the target output $\widehat{\mathbf{Y}}(t)$. We use linear regression and calculate the weights using the Moore-Penrose pseudo-inverse method [10]:

$$\mathbf{W}^{out'} = (\mathbf{X}'^T \cdot \mathbf{X}')^{-1} \cdot \mathbf{X}'^T \cdot \mathbf{Y}'. \tag{3}$$

Here, $\mathbf{W}^{out'}$ is the output weight vector extended with a the bias w_b, $\mathbf{X}'$ is the matrix of observation from the reservoir state where each row is represent the state of the reservoir at the corresponding time t and the columns represent the state of different nodes extended so that the last column is constant 1. Finally, $\widehat{\mathbf{Y}}'$ is the matrix of target output were each row represents the target output at the corresponding time t. Note

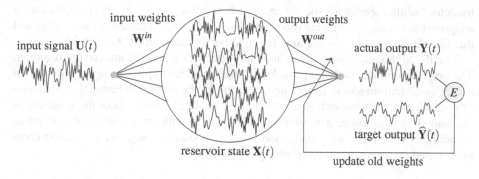

Fig. 2. Computation in a reservoir computer. The input signal $\mathbf{U}(t)$ is fed into every reservoir node i with a corresponding weight w_i^{in} denoted with weight column vector $\mathbf{W}^{in} = [w_i^{in}]$. Reservoir nodes are themselves coupled with each other using the weight matrix $\mathbf{W}^{res} = [w_{ij}^{res}]$, where w_{ij}^{res} is the weight of the connection from node j to node i.

that this also works for multi-dimensional output, in which case $\mathbf{W}^{out'}$ will be a matrix containing connection weights between each pair of reservoir nodes and output nodes.

Conceptually, the reservoir's role in RC is to act as a spatiotemporal kernel and project the input into a high-dimensional feature space [6]. In machine learning, this is usually referred to as feature extraction and is done to find hidden structures in data sets or time series. The output is then calculated by properly weighting and combining different features of the data [11]. An ideal reservoir should be able to perform feature extraction in a way that makes the mapping from feature space to output a linear problem. However, this is not always possible. In theory an ideal reservoir computer should have two features: a *separation property* of the reservoir and an *approximation property* of the readout layer. The former means the reservoir perturbations from two distinct inputs must remain distinguishable over time and the latter refers to the ability of the readout layer to map the reservoir state to a given target output in a sufficiently precise way.

Another way to understand computation in a high-dimensional recurrent systems is through analyzing their attractors. In this view, the state-space of the reservoir is partitioned into multiple basins of attraction. A basin of attraction is a subspace of the system's state-space, in which the system follows a trajectory towards its attractor. Thus computation takes place when the reservoir jumps between basins of attraction due to perturbations by an external input [12–15]. On the other hand, one could directly analyze computation in the reservoir as the reservoir's average instantaneous information content to produce a desired output [16].

There has been much research to find the optimal reservoir structure and readout strategy. Jaeger [17] suggests that in addition to the separation property, the reservoir should have fading memory to forget past inputs after some period of time. He achieves this by adjusting the standard deviation of the reservoir weight matrix σ so that the spectral radius of $\mathbf{W}^{res}$ remains close to 1, but slightly less than 1. This ensures that the reservoir can operate near critical dynamics, right at the edge between ordered and chaotic regimes. A key feature of critical systems is that perturbations to the system's trajectory neither spread nor die out, independent of the system size [18], which makes adaptive information processing robust to noise [14]. Other studies have also suggested that the critical dynamics is essential for good performance in RC [16, 19–22].

The RC architecture does not assume any specifics about the underlying reservoir. The only requirement is that it provides a suitable kernel to project inputs into a high-dimensional feature space. Reservoirs operating in the critical dynamical regime usually satisfy this requirement. Since RC makes no assumptions about the structure of the underlying reservoir, it is very suitable for use with unconventional computing paradigms [3–5]. Here, we propose and simulate a simple design for a reservoir computer based on a network of deoxyribozyme oscillators.

3   Reservoir Computing Using Deoxyribozyme Oscillators

To make a DNA reservoir computer, we first need a reservoir of DNA species with rich transient dynamics. To this end, we use a microfluidic reaction chamber in which different DNA species can interact. This must be an open reactor because we need to continually give input to the system and read its outputs. The reservoir state consists of the time-varying concentration of various species inside the chamber, and we compute using the reaction dynamics of the species inside the reactor. To perturb the reservoir we encode the time-varying input as fluctuations in the influx of species to the reactor. In [2, 23], a network of three deoxyribozyme NOT gates showed stable oscillatory dynamics in an open microfluidic reactor. We extend this work by designing a reservoir computer using deoxyribozyme-based oscillators and investigating their information-processing capabilities.

The oscillator dynamics in [2] suffices as an excitable reservoir. Ideally, the readout layer should also be implemented in a similar microfluidic setting and integrated with the reservoir. However, as a proof of concept we assume that we can read the reservoir state using fluorescent probes and calculate the output weights using software.

The oscillator network in [2] is described using a system of nine ordinary differential equations (ODEs), which simulate the details of a laboratory experiment of the proposed design. However, this model is mathematically unwieldy. We first reduce the oscillator ODEs in [2] to a form more amenable to mathematical analysis:

$$\frac{d[P_1]}{dt} = h\beta[S_1]([G_1] - [P_3]) - \frac{e}{V}[P_1], \quad \frac{d[S_1]}{dt} = \frac{S_1^m}{V} - h\beta[S_1]([G_1] - [P_3]) - \frac{e}{V}[S_1],$$

$$\frac{d[P_2]}{dt} = h\beta[S_2]([G_2] - [P_1]) - \frac{e}{V}[P_2], \quad \frac{d[S_2]}{dt} = \frac{S_2^m}{V} - h\beta[S_2]([G_2] - [P_1]) - \frac{e}{V}[S_2], \quad (4)$$

$$\frac{d[P_3]}{dt} = h\beta[S_3]([G_3] - [P_2]) - \frac{e}{V}[P_3], \quad \frac{d[S_3]}{dt} = \frac{S_3^m}{V} - h\beta[S_3]([G_3] - [P_2]) - \frac{e}{V}[S_3].$$

In this model, $[P_i]$, $[S_i]$, and $[G_i]$ are concentrations of three species of product molecules, three species of substrate molecules, and three species of gate molecules inside the reactor, and S_i^m is the influx rate of $[S_i]$. The brackets $[\cdot]$ indicate chemical concentration and should not be confused with the matrix notation introduced above. When explicitly talking about the concentrations at time t, we use $P_i(t)$ and $S_i(t)$. V is the volume of the reactor, h the fraction of the reactor chamber that is well-mixed, e is the efflux rate, and β is the reaction rate constant for the gate-substrate reaction, which is assumed to be identical for all gates and substrates, for simplicity.

To use this system as a reservoir we must ensure that it has transient or sustained oscillation. This can be easily analyzed by forming the Jacobian of the system. Observing that all substrate concentrations reach an identical and constant value relative to their magnitude, we can focus on the dynamics of the product concentrations and write an approximation to the Jacobian of the system as follows:

$$\mathbf{J} = \begin{bmatrix} \frac{d[P_1]}{d[P_1]} & \frac{d[P_1]}{d[P_2]} & \frac{d[P_1]}{d[P_3]} \\ \frac{d[P_2]}{d[P_1]} & \frac{d[P_2]}{d[P_2]} & \frac{d[P_2]}{d[P_3]} \\ \frac{d[P_3]}{d[P_1]} & \frac{d[P_3]}{d[P_2]} & \frac{d[P_3]}{d[P_3]} \end{bmatrix} = \begin{bmatrix} -\frac{e}{V} & -h\beta[S_1] & 0 \\ 0 & -\frac{e}{V} & -h\beta[S_2] \\ -h\beta[S_1] & 0 & -\frac{e}{V} \end{bmatrix} \tag{5}$$

Assuming that volume of the reactor V and the reaction rate constant β are given, the Jacobian is a function of only the efflux rate e and the substrate concentrations $[S_i]$. The eigenvalues of the Jacobian are given by:

$$\begin{aligned} \lambda_1 &= -h\beta([S_1][S_2][S_3])^{\frac{1}{3}} - \frac{e}{V} \\ \lambda_2 &= \frac{1}{2}h\beta([S_1][S_2][S_3])^{\frac{1}{3}} - \frac{e}{V} + \frac{\sqrt{3}}{2}h\beta([S_1][S_2][S_3])^{\frac{1}{3}}i \\ \lambda_3 &= \frac{1}{2}h\beta([S_1][S_2][S_3])^{\frac{1}{3}} - \frac{e}{V} - \frac{\sqrt{3}}{2}h\beta([S_1][S_2][S_3])^{\frac{1}{3}}i \end{aligned} \tag{6}$$

The existence of complex eigenvalues tells us that the system has oscillatory behavior near its critical points. The period of this oscillation is given by $T = 2\pi\frac{\sqrt{3}}{2}h\beta([S_1][S_2][S_3])^{-\frac{1}{3}}$ and can be adjusted by setting appropriate base values for S_i^m. For sustained oscillation, the real part of the eigenvalues should be zero, which can be obtained by a combination of efflux rate and substrate influx rates such that $\frac{1}{2}h\beta([S_1][S_2][S_3])^{\frac{1}{3}} - \frac{e}{V} = 0$.

This model works as follows. The substrate molecules enter the reaction chamber and are bound to and cleaved by active gate molecules that are immobilized inside the reaction chamber, e.g., on beads. This reaction turns substrate molecules into the corresponding product molecule. However, the presence of each product molecule concentration suppresses the reaction of other substrates and gates. These three coupled reaction and inhibition cycles give rise to the oscillatory behavior of the products' concentrations (Figure 4). Input is given to the system as fluctuation to one or more of the substrate influx rates. In Figure 3a we see that the concentration of S_1 varies rapidly as a response to the random fluctuations in S_1^m. This will result in antisymmetric concentrations of the substrate species inside the chamber and thus irregular oscillation of the concentration of product molecules. This irregular oscillation embeds features of the input fluctuation within it (Figure 3a). To keep the volume fixed, there is a continuous efflux of the chamber content. The Equation 4 assumes $([G_i] - [P_j]) > 0$, which should be taken into account while choosing initial concentrations and constants to simulate the system.

To perturb the intrinsic dynamics inside the reactor, an input signal can modulate one or more substrate influx rates. In our system, we achieve this by fluctuating S_1^m. In order to let the oscillators react to different values of S_i^m, we keep each new value of S_i^m

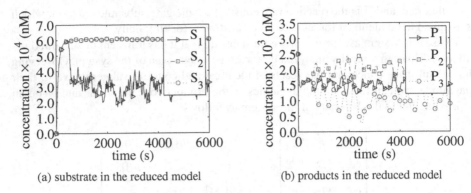

(a) substrate in the reduced model   (b) products in the reduced model

Fig. 3. The random fluctuation in substrate influx rate S_1^m leaves traces on the oscillator dynamics that can be read off a readout layer. We observe the traces of the substate influx rate fluctuation both in the dynamics of the substrate concentrations (a) and the product concentrations (b). Both substrate and product concentrations potentially carry information about the input. Substrate concentration S_1 is directly affected by S_1^m and therefore shows very rapid fluctuations.

constant for τ seconds. In a basic setup, the initial concentrations of all the substrates inside the reactor are zero. Two of the product concentrations $P_2(0)$ and $P_3(0)$ are also set to zero, but to break the symmetry in the system and let the oscillation begin we set $P_1(0) = 1000$ nM. The gate concentrations are set uniformly to $[G_i] = 2500$ nM. This ensures that $([G_i] - [P_j]) > 0$ in our setup. The base values for substrate-influx rates are set to 5.45×10^{-6} nmol s^{-1}. Figure 3 shows the traces of computer simulation of this model, where $\tau = 30$ s. We use the reaction rate constant from [2], $\beta = 5 \times 10^{-7}$ nM s^{-1}. Although the kinetics of immobilized deoxyribozyme may be different, for simplicity we use the reaction rate constant of free deoxyribozymes and we assume that we can find deoxyribozymes with appropriate kinetics when immobilized. The values for the remaining constants are $e = 8.8750 \times 10^{-2}$ nL s^{-1} and $h = 0.7849$, i.e., the average fraction of well-mixed solution calculated in [2]. We assume the same microscale continuous stirred-tank reactor (μCSTR) as [2, 23, 24], which has volume $V = 7.54$ nL. The small volume of the reactor lets us achieve high concentration of oligonucleotides with small amounts of material; a suitable experimental setup is described in [25].

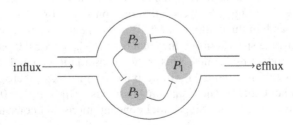

Fig. 4. Three products form an inhibitory cycle that leads to oscillatory behavior in the reservoir. Each product P_i inhibits the production of P_{i+1} by the corresponding deoxyribozyme (cf. Equation 4).

The dynamics of the substrates (Figure 3a) and products (Figure 3a) are instructive as to what we can use as our reservoir state. Our focus will be the product concentrations. However, the substrate concentrations also show interesting irregular behavior that can potentially carry information about the input signals. This is not surprising since all of the substrate and product concentrations are connected in our oscillator network. However, the one substrate that is directly affected by the influx (S_1 in this case) shows the most intense fluctuations that are directly correlated with the input. In some cases providing this extra information to the readout layer can help to find the right mapping between the reservoir state and the target output.

In the next section, we build two different reservoir computers using the dynamics of the concentrations in the reactor and use them to solve sample temporal tasks. Despite the simplicity of our system, it can retain the memory of past inputs inside the reservoir and use it to produce output.

4   Task Solving Using a Deoxyribozyme Reservoir Computer

We saw in the preceding section that we can use substrate influx fluctuation as input to our molecular reservoir. We now show that we can train a readout layer to map the dynamics of the oscillator network to a target output. Recall that τ is the input hold time during which we keep S_1^m constant so that the oscillators can react to different values of S_1^m. In other words, at the beginning of each τ interval a new random value for substrate influx is chosen and held fixed for τ seconds. Here, we set the input hold time $\tau = 100$ s. In addition, before computing with the reservoir we must make sure that it has settled in its natural dynamics, otherwise the output layer will see dynamical behavior that is due to the initial conditions of the oscillators and not the input provided to the system. In the model, the oscillators reach their stable oscillation pattern within 500 s. Therefore, we start our reservoir by using a fixed S_1^m as described in Section 3 and run it for 500 s before introducing fluctuations in S_1^m.

To study the performance of our DNA reservoir computer we use two different tasks, Task A and Task B, as toy problems. Both have recursive time dependence and therefore require the reservoir to remember past inputs for some period of time, and both are simplified versions of a popular RC benchmark, NARMA [8]. We define the input as $S_1^m(t) = S_1^{m*} R$, where S_1^{m*} is the influx rate used for the normal working of the oscillators (5.45×10^{-6} nmol s^{-1} in our experiment) and R is a random value between 0 and 1 sampled from a uniform distribution. We define the target output $\widehat{\mathbf{Y}}(\mathbf{t})$ of Task A as follows:

$$\widehat{\mathbf{Y}}(t) = S_1^m(t-1) + 2S_1^m(t-2). \tag{7}$$

For Task B, we increase the length of the time dependence and make it a function of input hold time τ. We define the target output as follows:

$$\widehat{\mathbf{Y}}(t) = S_1^m(t-\tau) + \frac{1}{2}S_1^m(t - \frac{3}{2}\tau). \tag{8}$$

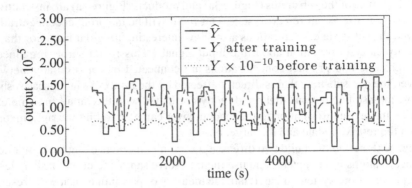

Fig. 5. Target output and the output of the molecular reservoir computer on Task A (Equation 7) before and after training. After 500 s the input starts to fluctuate randomly every τ seconds. In this example, the output of the system before training is 10 orders of magnitude larger than the target output. We rescaled the output before training to be able to show it in this plot. After training, the output is in the range of the target output and it tracks the fluctuations in the target output more closely.

Note that the vectors $\widehat{\mathbf{Y}}(t)$ and $\mathbf{Y}(t)$ have only one row in this example. Figure 5 shows an example of the reservoir output $\mathbf{Y}(t)$ and the target output $\widehat{\mathbf{Y}}(t)$ calculated using Equation 7 before and after training. In this example, the reservoir output before training is 10 orders of magnitude off the target.

Our goal is to find a set of output weights so that $\mathbf{Y}(t)$ tracks the target output as closely as possible. We calculate the error using normalized root-mean-square error (NRMSE) as follows:

$$\text{NRMSE} = \frac{1}{Y_{max} - Y_{min}} \sqrt{\frac{\sum_{t=t_1}^{t_n} (\widehat{\mathbf{Y}}(t) - \mathbf{Y}(t))^2}{n}}, \qquad (9)$$

where Y_{max} and Y_{min} are the maximum and the minimum of the $\mathbf{Y}(t)$ during the time interval $t_1 < t < t_n$. The denominator $Y_{max} - Y_{min}$ is to ensure that $0 \leq \text{NRMSE} \leq 1$, where $\text{NRMSE} = 0$ means $\mathbf{Y}(t)$ matches $\widehat{\mathbf{Y}}(t)$ perfectly.

Now we propose two different ways of calculating the output from the reservoir: (1) using only the dynamics of the product concentrations and (2) using both the product and substrate concentrations. To formalize this using the block matrix notation, for the product-only version the reservoir state is given by $\mathbf{X}(t) = \mathbf{P}(t) = [P_1(t)\ P_2(t)\ P_3(t)]^T$. For the product-and-substrate version the reservoir state is given by vertically appending $\mathbf{S}(t) = [S_1(t)\ S_2(t)\ S_3(t)]^T$ to $\mathbf{P}(t)$, i.e., $\mathbf{X}(t) = [\mathbf{P}(t)\ \mathbf{S}(t)]^T$, where $\mathbf{P}(t)$ is the column vector of the product concentrations as before and $\mathbf{S}(t)$ is the column vector of the substrate concentrations. We use 2000 s of the reservoir dynamics $\mathbf{X}(t)$ to calculate the output weight matrix $\mathbf{W}^{out}$ using linear regression. We then test the generalization, i.e., how well the output $\mathbf{Y}(t)$ tracks the target $\widehat{\mathbf{Y}}(t)$ during another 2000 s period that we did not use to calculate the weights.

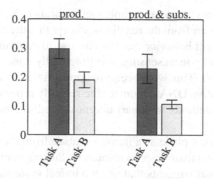

Fig. 6. Generalization NRMSE of the product-only and the product-and-substrate molecular reservoir computer on Task A (Equation 7) and Task B (Equation 8) averaged over 100 trials. The bars and error bars show the mean and the standard deviation of NRMSE respectively.

Figure 6 shows the mean and standard deviation of NRMSE of the reservoir computer using two different readout layer solving Task A and Task B. The product-and-substrate reservoir achieves a mean NRMSE of 0.23 and 0.11 on Task A and Task B with standard deviations 0.05 and 0.02 respectively, and the product-only reservoir achieves a mean NRMSE of 0.30 and 0.19 on Task A and Task B with standard deviations 0.04 and 0.03 respectively. As expected, the product-and-substrate reservoir computer achieves about 10% improvement over the product-only version owing to its higher phase space dimensionality. Furthermore, both reservoirs achieve a 10% improvement on Task B over Task A. This is surprising at first because Task B requires the reservoir to remember the input over a time interval of $\frac{3}{2}\tau$, but Task A only requires the last two time steps. However, to extract the features in the input signal, the input needs to percolate in the reservoir, which takes more than just two time steps. Task B requires more memory of the input, but also gives the reservoir enough time to process the input signal, which results in higher performance. Similar effects have been observed in [16]. Therefore, despite the very simple reservoir structure (three coupled oscillators), we can compute simple temporal tasks with 90% accuracy. Increasing the number of oscillators and using the history of the oscillators dynamics similar to [26] could potentially lead to even higher performance.

5 Discussion and Related Work

DNA chemistry is inherently programmable and highly versatile, and a number of different techniques have been developed, such as building digital and analog circuits using strand displacement cascades [27, 28], developing game-playing molecular automata using deoxyribozymes [29], and directing self-assembly of nanostructures   [30–32]. All of these approaches require precise design of DNA sequences to form the required structures and perform the desired computation. In this paper, we proposed a reservoir-computing approach to molecular computing. In nature, evidence for reservoir computing has been found in systems as simple as a bucket of water [33], simple organisms such as *E. Coli* [34], and in systems as complex as the brain [35]. This approach does not require any specific behavior from the reactions, except that the reaction dynamics must result in a suitable transient behavior that we can use to compute [6]. This could give us a new perspective in long-term sensing, and potentially controlling, gene expression patterns over time in a cell. This would require appropriate sensors to detect cell state, for example the pH-sensitive DNA nanomachine recently reported by Modi et al. [36]. This may result in new methods for smart diagnosis and treatment using DNA signal translators [37–39].

In RC, computation takes place as a transformation from the input space to a high-dimensional spatiotemporal feature space created by the transient dynamics of the reservoir. Mathematical analysis suggests that all dynamical systems show the same information processing capacity [40]. However, in practice, the performance of a reservoir is significantly affected by its dynamical regime. Many studies have shown that to achieve a suitable reservoir in general, the underlying dynamical system must operate in the critical dynamical regime [8, 16, 20, 21].

We used the dynamics of the concentrations of different molecular species to extract features of an input signal and map them to a desired output. As a proof of concept, we proposed a reservoir computer using deoxyribozyme oscillator network and showed how to provide it with input and read its outputs. However, in our setup, we assumed that we read the reservoir state using fluorescent probes and process them using software. In principle, the mapping from the reservoir state to target output can be carried out as an integrated part of the chemistry using an approach similar to the one reported in [28], which implements a neural network using strand displacement. In [41], we proposed a chemical reaction network inspired by deoxyribozyme chemistry that can learn a linear function and repeatedly use it to classify input signals. In principle, these methods could be used to implement the regression algorithm and therefore the readout layer as an integrated part of the molecular reservoir computer. A microfluidic reactor has been demonstrated in [25] that would be suitable for implementing our system. Therefore, the molecular reservoir computer that we proposed here is physically plausible and can be implemented in the laboratory using microfuidics.

6   Conclusion and Future Work

We have proposed and simulated a novel approach to DNA computing based on the reservoir computing paradigm. Using a network of oscillators built from deoxyribozymes we can extract hidden features in a given input signal and compute any desired output. We tested the performance of this approach on two simple temporal tasks. This approach is generalizable to different molecular species so long as they possess rich reaction dynamics. Given the available technology today this approach is plausible and can lead to many innovations in biological signal processing, which has important applications in smart diagnosis and treatment techniques. In future work, we shall study the use of other sets of reactions for the reservoir. Moreover, for any real-world application of this technique, we have to address the chemical implementation of the readout layer. An important open question is the complexity of molecular reactions necessary to achieve critical dynamics in the reservoir. For practical applications, the effect of sparse input and sparse readout needs thorough investigation, i.e., how should one distribute the input to the reservoir and how much of the reservoir dynamics is needed for the readout layer to reconstruct the target output accurately? It is also possible to use the history of the reservoir dynamics to compute the output, which would require addition of a feedback channel to the reactor. The molecular readout layer could be set up to read the species concentration along the feedback channel. Another possibility is to connect many reactors to create a modular molecular reservoir computer, which could be used strategically to scale up to more complex problems.

References

1. Jaeger, H., Haas, H.: Harnessing nonlinearity: Predicting chaotic systems and saving energy in wireless communication. Science 304(5667), 78–80 (2004)
2. Farfel, J., Stefanovic, D.: Towards practical biomolecular computers using microfluidic deoxyribozyme logic gate networks. In: Carbone, A., Pierce, N.A. (eds.) DNA11. LNCS, vol. 3892, pp. 38–54. Springer, Heidelberg (2006)
3. Lukoševičius, M., Jaeger, H., Schrauwen, B.: Reservoir computing trends. KI - Künstliche Intelligenz 26(4), 365–371 (2012)
4. Smerieri, A., Duport, F., Paquot, Y., Schrauwen, B., Haelterman, M., Massar, S.: Analog readout for optical reservoir computers. In: Bartlett, P., Pereira, F., Burges, C., Bottou, L., Weinberger, K. (eds.) Advances in Neural Information Processing Systems 25: 26th Annual Conference on Neural Information Processing Systems, pp. 953–961. Curran Associates, Inc (2012)
5. Paquot, Y., Duport, F., Smerieri, A., Dambre, J., Schrauwen, B., Haelterman, M., Massar, S.: Optoelectronic reservoir computing. Scientific Reports 2 (2012)
6. Lukoševičius, M., Jaeger, H.: Reservoir computing approaches to recurrent neural network training. Computer Science Review 3(3), 127–149 (2009)
7. Maass, W., Natschläger, T., Markram, H.: Real-time computing without stable states: a new framework for neural computation based on perturbations. Neural Computation 14(11), 2531–2560 (2002)
8. Jaeger, H.: Tutorial on training recurrent neural networks, covering BPPT, RTRL, EKF and the "echo state network" approach. Technical Report GMD Report 159, German National Research Center for Information Technology, St. Augustin-Germany (2002)
9. Widrow, B., Lehr, M.: 30 years of adaptive neural networks: Perceptron, madaline, and back-propagation. Proceedings of the IEEE 78(9), 1415–1442 (1990)
10. Penrose, R.: A generalized inverse for matrices. Mathematical Proceedings of the Cambridge Philosophical Society 51, 406–413 (1955)
11. Bishop, C.M.: Pattern Recognition and Machine Learning (Information Science and Statistics). Springer-Verlag New York, Inc., Secaucus (2006)
12. Sussillo, D., Barak, O.: Opening the black box: Low-dimensional dynamics in high-dimensional recurrent neural networks. Neural Computation 25(3), 626–649 (2012)
13. Sussillo, D., Abbott, L.F.: Generating coherent patterns of activity from chaotic neural networks. Neuron 63(4), 544–557 (2009)
14. Goudarzi, A., Teuscher, C., Gulbahce, N., Rohlf, T.: Emergent criticality through adaptive information processing in Boolean networks. Phys. Rev. Lett. 108, 128702 (2012)

15. Krawitz, P., Shmulevich, I.: Basin entropy in boolean network ensembles. Phys. Rev. Lett. 98(15), 158701 (2007)
16. Snyder, D., Goudarzi, A., Teuscher, C.: Computational capabilities of random automata networks for reservoir computing. Phys. Rev. E 87, 042808 (2013)
17. Jaeger, H.: Short term memory in echo state networks. Technical Report GMD Report 152, GMD-Forschungszentrum Informationstechnik (2002)
18. Rohlf, T., Gulbahce, N., Teuscher, C.: Damage spreading and criticality in finite random dynamical networks. Phys. Rev. Lett. 99(24), 248701 (2007)
19. Natschläger, T., Maass, W.: Information dynamics and emergent computation in recurrent circuits of spiking neurons. In: Thrun, S., Saul, L., Schoelkpf, B. (eds.) Proc. of NIPS 2003, Advances in Neural Information Processing Systems, vol. 16, pp. 1255–1262. MIT Press, Cambridge (2004)
20. Bertschinger, N., Natschläger, T.: Real-time computation at the edge of chaos in recurrent neural networks. Neural Computation 16(7), 1413–1436 (2004)
21. Büsing, L., Schrauwen, B., Legenstein, R.: Connectivity, dynamics, and memory in reservoir computing with binary and analog neurons. Neural Computation 22(5), 1272–1311 (2010)
22. Boedecker, J., Obst, O., Mayer, N.M., Asada, M.: Initialization and self-organized optimization of recurrent neural network connectivity. HFSP Journal 3(5), 340–349 (2009)
23. Morgan, C., Stefanovic, D., Moore, C., Stojanovic, M.N.: Building the components for a biomolecular computer. In: Ferretti, C., Mauri, G., Zandron, C. (eds.) DNA10. LNCS, vol. 3384, pp. 247–257. Springer, Heidelberg (2005)
24. Chou, H.P., Unger, M., Quake, S.: A microfabricated rotary pump. Biomedical Microdevices 3(4), 323–330 (2001)
25. Galas, J.C., Haghiri-Gosnet, A.M., Estevez-Torres, A.: A nanoliter-scale open chemical reactor. Lab Chip 13, 415–423 (2013)
26. Appeltant, L., Soriano, M.C., Van der Sande, G., Danckaert, J., Massar, S., Dambre, J., Schrauwen, B., Mirasso, C.R., Fischer, I.: Information processing using a single dynamical node as complex system. Nature Communications 2 (2011)
27. Qian, L., Winfree, E.: Scaling up digital circuit computation with DNA strand displacement cascades. Science 332(6034), 1196–1201 (2011)
28. Qian, L., Winfree, E., Bruck, J.: Neural network computation with DNA strand displacement cascades. Nature 475(7356), 368–372 (2011)
29. Pei, R., Matamoros, E., Liu, M., Stefanovic, D., Stojanovic, M.N.: Training a molecular automaton to play a game. Nature Nanotechnology 5(11), 773–777 (2010)
30. Yin, P., Choi, H.M.T., Calvert, C.R., Pierce, N.A.: Programming biomolecular self-assembly pathways. Nature 451(7176), 318–322 (2008)
31. Wei, B., Dai, M., Yin, P.: Complex shapes self-assembled from single-stranded DNA tiles. Nature 485(7400), 623–626 (2012)
32. Ke, Y., Ong, L.L., Shih, W.M., Yin, P.: Three-dimensional structures self-assembled from DNA bricks. Science 338(6111), 1177–1183 (2012)
33. Fernando, C., Sojakka, S.: Pattern recognition in a bucket. In: Banzhaf, W., Ziegler, J., Christaller, T., Dittrich, P., Kim, J.T. (eds.) ECAL 2003. LNCS (LNAI), vol. 2801, pp. 588–597. Springer, Heidelberg (2003)
34. Jones, B., Stekel, D., Rowe, J., Fernando, C.: Is there a liquid state machine in the bacterium *Escherichia coli*? In: IEEE Symposium on Artificial Life, ALIFE 2007, pp. 187–191 (2007)
35. Yamazaki, T., Tanaka, S.: The cerebellum as a liquid state machine. Neural Networks 20(3), 290–297 (2007)

36. Modi, S., Nizak, C., Surana, S., Halder, S., Krishnan, Y.: Two DNA nanomachines map pH changes along intersecting endocytic pathways inside the same cell. Nat. Nano 8(6), 459–467 (2013)
37. Beyer, S., Dittmer, W., Simmel, F.: Design variations for an aptamer-based DNA nanodevice. Journal of Biomedical Nanotechnology 1(1), 96–101 (2005)
38. Beyer, S., Simmel, F.C.: A modular DNA signal translator for the controlled release of a protein by an aptamer. Nucleic Acids Research 34(5), 1581–1587 (2006)
39. Shapiro, E., Gil, B.: RNA computing in a living cell. Science 322(5900), 387–388 (2008)
40. Dambre, J., Verstraeten, D., Schrauwen, B., Massar, S.: Information processing capacity of dynamical systems. Scientific Reports 2 (2012)
41. Lakin, M.R., Minnich, A., Lane, T., Stefanovic, D.: Towards a biomolecular learning machine. In: Durand-Lose, J., Jonoska, N. (eds.) UCNC 2012. LNCS, vol. 7445, pp. 152–163. Springer, Heidelberg (2012)

Signal Transmission across Tile Assemblies: 3D Static Tiles Simulate Active Self-assembly by 2D Signal-Passing Tiles

Jacob Hendricks[1,*], Jennifer E. Padilla[2,**],
Matthew J. Patitz[1,*], and Trent A. Rogers[3,*]

[1] Department of Computer Science and Computer Engineering,
University of Arkansas
{jhendric,patitz}@uark.edu
[2] Dept. of Chem., New York U
jp164@nyu.edu
[3] Department of Mathematical Sciences, University of Arkansas
tar003@email.uark.edu

Abstract. The 2-Handed Assembly Model (2HAM) is a tile-based self-assembly model in which, typically beginning from single tiles, arbitrarily large aggregations of static tiles combine in pairs to form structures. The Signal-passing Tile Assembly Model (STAM) is an extension of the 2HAM in which the tiles are dynamically changing components which are able to alter their binding domains as they bind together. In this paper, we prove that there exists a 3D tile set in the 2HAM which is intrinsically universal for the class of all 2D STAM$^+$ systems at temperature 1 and 2 (where the STAM$^+$ does not make use of the STAM's power of glue deactivation and assembly breaking, as the tile components of the 2HAM are static and unable to change or break bonds). This means that there is a single tile set U in the 3D 2HAM which can, for an arbitrarily complex STAM$^+$ system S, be configured with a single input configuration which causes U to exactly simulate S at a scale factor dependent upon S. Furthermore, this simulation uses only 2 planes of the third dimension.

To achieve this result, we also demonstrate useful techniques and transformations for converting an arbitrarily complex STAM$^+$ tile set into an STAM$^+$ tile set where every tile has a constant, low amount of complexity, in terms of the number and types of "signals" they can send, with a trade off in scale factor.

While the first result is of more theoretical interest, showing the power of static tiles to simulate dynamic tiles when given one extra plane in 3D, the second is of more practical interest for the experimental implementation of STAM tiles, since it provides potentially useful strategies for developing powerful STAM systems while keeping the complexity of individual tiles low, thus making them easier to physically implement.

---

* Supported in part by National Science Foundation Grant CCF-1117672.
** This author's research was supported by National Science Foundation Grant CCF-1117210.

D. Soloveichik and B. Yurke (Eds.): DNA 2013, LNCS 8141, pp. 90–104, 2013.

1   Introduction

Self-assembling systems are those in which large, disorganized collections of relatively simple components autonomously, without external guidance, combine to form organized structures. Self assembly drives the formation of a vast multitude of naturally forming structures, across a wide range of sizes and complexities (from the crystalline structure of snowflakes to complex biological structures such as viruses). Recognizing the immense power and potential of self-assembly to manufacture structures with precision down to the molecular level, researchers have been pursuing the creation and study of artificial self-assembling systems. This research has led to steadily increasing sophistication of both the theoretical models (from the Tile Assembly Model (TAM) [21], to the 2-Handed Assembly Model (2HAM) [4, 8], and many others [1–3, 8, 12]) as well as experimentally produced building blocks and systems (a mere few of which include [5, 13, 15, 16, 19, 20]). While a number of models exist for passive self-assembly, as can be seen above, research into modeling active self-assembly is just beginning [18, 22]. Unlike passive self-assembly where structures bind and remain in one state, active self-assembly allows for structures to bind and then change state.

A newly developed model, the Signal-passing Tile Assembly Model (STAM) [18], is based upon the 2HAM but with a powerful and important difference. Tiles in the aTAM and 2HAM are static, unchanging building blocks which can be thought of as analogous to write-once memory, where a location can change from empty to a particular value once and then never change again. Instead, the tiles of the STAM each have the ability to undergo some bounded number of transformations as they bind to an assembly and while they are connected. Each transformation is initiated by the binding event of a tile's glue, and consists of some other glue on that tile being turned either "on" or "off". By chaining together sequences of such events which propagate across the tiles of an assembly, it is possible to send "signals" which allow the assembly to adapt during growth. Since the number of transitions that any glue can make is bounded, this doesn't provide for "fully reusable" memory, but even with the limited reuse it has been shown that the STAM is more powerful than static models such as the aTAM and 2HAM (in 2D), for instance being able to strictly self-assemble the Sierpinski triangle [18]. A very important feature of the STAM is its asynchronous nature, meaning that there is no timeframe during which signals are guaranteed to fully propagate, and no guaranteed ordering to the arrival of multiple signals. Besides providing a useful theoretical framework of asynchronous behavior, the design of the STAM was carefully aligned to the physical reality of implementation by DNA tiles using cascades of strand-displacement. Capabilities in this area are improving, and now include the linear transmission of signals, where one glue binding event can activate one other glue on a DNA tile [17].

Although the STAM is intended to provide both a powerful theoretical framework and a solid basis for representing possible physical implementations, often those two goals are at odds. In fact, in the STAM it is possible to define tiles which have arbitrary *signal complexity* in terms of the numbers of glues that they have on any given side and the number of signals that each tile can initiate. Clearly, with increasing complexity of individual tiles, the ease of making them in the laboratory diminishes. Therefore, in this paper our first set of results provide a variety of methods for simplifying the tiles in STAM systems. Besides reducing just the general signal complexity of tiles, we also seek to reduce and/or remove certain patterns of signals which may be more difficult to build into DNA-based tiles, namely *fan-out* (which occurs when a single signal must split into multiple paths and have multiple destinations), *fan-in* (which occurs when multiple signals must converge and join into one path to arrive at a single glue), and *mutual activation* (which occurs when both of the glues participating in a particular binding event initiate their own signals). By trading signal complexity for tile complexity and scale factor, we show how to use some simple primitive substitutions to reduce STAM tile sets to those with much simpler tiles. Note that while in the general STAM it is possible for signals to turn glues both "on" and "off", our results pertain only to systems which turn glues "on" (which we call $STAM^+$ systems).

In particular, we show that the tile set for any temperature 1 $STAM^+$ system, with tiles of arbitrary complexity, can be converted into a temperature 1 $STAM^+$ system with a tile set where no tile has greater then 2 signals and either fan-out or mutual activation are completely eliminated. We show that any temperature 2 $STAM^+$ system can be converted into a temperature 2 $STAM^+$ system where no tile has greater than 1 signal and both fan-out and mutual activation are eliminated. Importantly, while both conversions have a worst case scale factor of $|T^2|$, where T is the tile set of the original system, and worst case tile complexity of $|T^2|$, those bounds are required for the extremely unrealistic case where *every* glue is on *every* edge of some tile and also sends signals to *every* glue on *every* side of that tile. Converting from a more realistic tile set yields factors which are on the order of the square of the maximum signal complexity for each side of a tile, which is typically much smaller. Further, the techniques used to reduce signal complexity and remove fan-out and mutual activation are likely to be useful in the original design of tile sets rather than just as brute force conversions of completed tile sets.

We next consider the topic of intrinsic universality, which was initially developed to aid in the study of cellular automata [6, 7]. The notion of intrinsic universality was designed to capture a strong notion of simulation, in which one particular automaton is capable of simulating the *behavior* of any automaton within a class of automata. Furthermore, to simulate the behavior of another automaton, the simulating automaton must evolve in such a way that a translated rescaling (rescaled not only with respect to rectangular blocks of cells, but also with respect to time) of the simulator can be mapped to a configuration of the simulated automaton. The specific rescaling depends on the simulated automaton and gives rise to a global rule such that each step of the simulated automaton's evolution is mirrored by the simulating automaton, and vice versa via the inverse of the rule.

In this way, it is said that the simulator captures the dynamics of the simulated system, acting exactly like it, modulo scaling. This is in contrast to a computational simulation, for example when a general purpose digital computer runs a program to simulate a cellular automata while the processor's components don't actually arrange themselves as, and behave like, a grid of cellular automata. In [11], it was shown that the aTAM is intrinsically universal, which means that there is a single tile set U such that, for any aTAM tile assembly system $\mathcal{T}$ (of any temperature), the tiles of U can be arranged into a seed structure dependent upon $\mathcal{T}$ so that the resulting system (at temperature 2), using only the tiles from U, will faithfully simulate the behaviors of $\mathcal{T}$. In contrast, in [9] it was shown that no such tile set exists for the 2HAM since, for every temperature, there is a 2HAM system which cannot be simulated by any system operating at a lower temperature. Thus no tile set is sufficient to simulate 2HAM systems of arbitrary temperature.

For our main result, we show that there is a 3D 2HAM tile set U which is intrinsically universal (IU) for the class $\mathfrak{C}$ of all STAM$^+$ systems at temperature 1 and 2. For every $\mathcal{T} \in \mathfrak{C}$, a single input supertile can be created, and using just copies of that input supertile and the tiles from U, at temperature 2 the resulting system with faithfully simulate $\mathcal{T}$. Furthermore, the simulating system will use only 2 planes of the third dimension. (The signal tile set simplification results are integral in the construction for this result, especially in allowing it to use only 2 planes.) This result is noteworthy especially because it shows that the dynamic behavior of signal tiles (excluding glue deactivation) can be *fully duplicated* by static tile systems which are allowed to "barely" use three dimensions. Furthermore, for every temperature $\tau > 1$ there exists a 3D 2HAM tile set which can simulate the class of all STAM$^+$ systems at temperature τ.

2   Preliminaries

Here we provide informal descriptions of the models and terms used in this paper. Due to space limitations, the formal definitions can be found in [14].

2.1   Informal Definition of the 2HAM

The 2HAM [4,8] is a generalization of the abstract Tile Assembly Model (aTAM) [21] in that it allows for two assemblies, both possibly consisting of more than one tile, to attach to each other. Since we must allow that the assemblies might require translation before they can bind, we define a *supertile* to be the set of all translations of a τ-stable assembly, and speak of the attachment of supertiles to each other, modeling that the assemblies attach, if possible, after appropriate translation. We now give a brief, informal, sketch of the d-dimensional 2HAM, for $d \in \{2, 3\}$, which is normally defined as a 2D model but which we extend to 3D as well, in the natural and intuitive way.

A *tile type* is a unit square if $d = 2$, and cube if $d = 3$, with each side having a *glue* consisting of a *label* (a finite string) and *strength* (a non-negative integer). We assume a finite set T of tile types, but an infinite number of copies of each tile type, each copy referred to as a *tile*. A *supertile* is (the set of all translations of) a positioning of tiles on the integer lattice $\mathbb{Z}^d$. Two adjacent tiles in a supertile *interact* if the glues on their abutting sides are equal and have positive strength. Each supertile induces a *binding graph*, a grid graph whose vertices are tiles, with an edge between two tiles if they interact. The supertile is τ-*stable* if every cut of its binding graph has strength at least τ, where the weight of an edge is the strength of the glue it represents. That is, the supertile is stable if at least energy τ is required to separate the supertile into two parts. A 2HAM *tile assembly system* (TAS) is a pair $\mathcal{T} = (T, \tau)$, where T is a finite tile set and τ is the *temperature*, usually 1 or 2. (Note that this is considered the "default" type of 2HAM system, while a system can also be defined as a triple (T, S, τ), where S is the *initial configuration* which in the default case is just infinite copies of all tiles from T, but in other cases can additionally or instead consist of copies of pre-formed supertiles.) Given a TAS $\mathcal{T} = (T, \tau)$, a supertile is *producible*, written as $\alpha \in \mathcal{A}[\mathcal{T}]$, if either it is a single tile from T, or it is the τ-stable result of translating two producible assemblies without overlap. Note that if $d = 3$, or if $d = 2$ but it is explicitly mentioned that *planarity* is to be preserved, it must be possible for one of the assemblies to start infinitely far from the other and by merely translating in d dimensions arrive into a position such that the combination of the two is τ-stable, without ever requiring overlap. This prevents, for example, binding on the interior of a region completely enclosed by a supertile. A supertile α is *terminal*, written as $\alpha \in \mathcal{A}_\square[\mathcal{T}]$, if for every producible supertile β, α and β cannot be τ-stably attached. A TAS is *directed* if it has only one terminal, producible supertile.

2.2   Informal Description of the STAM

In the STAM, tiles are allowed to have sets of glues on each edge (as opposed to only one glue per side as in the TAM and 2HAM). Tiles have an initial state in which each glue is either "on" or "latent" (i.e. can be switched on later). Tiles also each implement a transition function which is executed upon the binding of any glue on any edge of that tile. The transition function specifies, for each glue g on a tile, a set of glues (along with the sides on which those glues are located) and an action, or *signal* which is *fired* by g's binding, for each glue in the set. The actions specified may be to: 1. turn the glue on (only valid if it is currently latent), or 2. turn the glue off (valid if it is currently on or latent). This means that glues can only be on once (although may remain so for an arbitrary amount of time or permanently), either by starting in that state or being switched on from latent (which we call *activation*), and if they are ever switched to off (called *deactivation*) then no further transitions are allowed for that glue. This essentially provides a single "use" of a glue (and the signal sent by its binding). Note that turning a glue off breaks any bond that that glue may have formed with a neighboring tile. Also, since tile edges can have multiple active glues, when tile edges with multiple glues are adjacent, it is assumed that all matching glues in the on state bind (for a total binding strength equal to the sum of the strengths of the individually bound glues). The transition function defined for each tile type is allowed a unique set of output actions for the binding event of each glue along its edges, meaning that the binding of any particular glue on a tile's edge can initiate a set of actions to turn an arbitrary set of the glues on the sides of the same tile either on or off.

As the STAM is an extension of the 2HAM, binding and breaking can occur between tiles contained in pairs of arbitrarily sized supertiles. In order to allow for physical mechanisms which implement the transition functions of tiles but are arbitrarily slower or faster than the average rates of (super)tile attachments and detachments, rather than immediately enacting the outputs of transition functions, each output action is put into a set of "pending actions" which includes all actions which have not yet been enacted for that glue (since it is technically possible for more than one action to have been initiated, but not yet enacted, for a particular glue). Any event can be randomly selected from the set, regardless of the order of arrival in the set, and the ordering of either selecting some action from the set or the combination of two supertiles is also completely arbitrary. This provides fully asynchronous timing between the initiation, or firing, of signals (i.e. the execution of the transition function which puts them in the pending set) and their execution (i.e. the changing of the state of the target glue), as an arbitrary number of supertile binding events may occur before any signal is executed from the pending set, and vice versa.

An STAM system consists of a set of tiles and a temperature value. To define what is producible from such a system, we use a recursive definition of producible assemblies which starts with the initial tiles and then contains any supertiles which can be formed by doing the following to any producible assembly: 1. executing any entry from the pending actions of any one glue within a tile within that supertile (and then that action is removed from the pending set), 2. binding with another supertile if they are able to form a τ-stable supertile, or 3. breaking into 2 separate supertiles along a cut whose total strength is $< \tau$.

The STAM, as formulated, is intended to provide a model based on experimentally plausible mechanisms for glue activation and deactivation. However, while the model allows for the placement of an arbitrary number of glues on each tile side and for each of them to signal an arbitrary number of glues on the same tile, this is (currently quite) limited in practice. Therefore, each system can be defined to take into account a desired threshold for each of those parameters, not exceeding it for any given tile type, and so we have defined the notion of *full-tile signal complexity* as the maximum number of signals on any tile in a set (see [14]) to capture the maximum complexity of any tile in a given set.

Definition 1. *We define the STAM$^+$ to be the STAM restricted to using only glue activation, and no glue deactivation. Similarly, we say an STAM$^+$ tile set is one which contains no defined glue deactivation transitions, and an STAM$^+$ system $\mathcal{T} = (T, \tau)$ is one in which T is an STAM$^+$ tile set.*

As the main goal of this paper is to show that self-assembly by systems using active, signalling tiles can be simulated using the static, unchanging tiles of the 3D 2HAM, since they have no ability to break apart after forming τ-stable structures, all of our results are confined to the STAM$^+$.

A detailed, technical definition of the STAM model is provided in [14].

2.3 Informal Definitions for Simulation

Here we informally describe what it means for one 2HAM or STAM TAS to "simulate" another. Formal definitions, adapted from those of [9], can be found in [14] .

Let $\mathcal{U} = (U, S_U, \tau_U)$ be the system which is simulating the system $\mathcal{T} = (T, S_T, \tau_T)$. There must be some scale factor $c \in \mathbb{N}$ at which $\mathcal{U}$ simulates $\mathcal{T}$, and we define a *representation function* R which maps each $c \times c$ square (sub)assembly in $\mathcal{U}$ to a tile in $\mathcal{T}$ (or empty space if it is incomplete). Each such $c \times c$ block is referred to as a *macrotile*, since that square configuration of tiles from set U represent a single tile from set T. We say that $\mathcal{U}$ simulates $\mathcal{T}$ under representation function R at scale c.

To properly simulate $\mathcal{T}$, $\mathcal{U}$ must have 1. *equivalent productions*, meaning that every supertile producible in $\mathcal{T}$ can be mapped via R to a supertile producible in $\mathcal{U}$, and vice versa, and 2. *equivalent dynamics*, meaning that when any two supertiles α and β, which are producible in $\mathcal{T}$, can combine to form supertile γ, then there are supertiles producible in $\mathcal{U}$ which are equivalent to α and β which can combine to form a supertile equivalent to γ, and vice versa. Note that especially the formal definitions for equivalent dynamics include several technicalities related to the fact that multiple supertiles in $\mathcal{U}$ may map to a single supertile in $\mathcal{T}$, among other issues. Please see [14] for details.

We say that a tile set U is *intrinsically universal* for a class of tile assembly systems if, for every system in that class, a system can be created for which 1. U is the tile set, 2. there is some initial configuration which consists of supertiles created from tiles in U, where those "input" supertiles are constructed to encode information about the system being simulated, and perhaps also singleton tiles from U, 3. a representation function which maps macrotiles in the simulator to tiles in the simulated system, and 4. under that representation function, the simulator has equivalent productions and equivalent dynamics to the simulated system. Essentially, there is one tile set which can simulate any system in the class, using only custom configured input supertiles.

3 Transforming STAM$^+$ Systems from Arbitrary to Bounded Signal Complexity

In this section, we demonstrate methods for reducing the signal complexity of STAM$^+$ systems with $\tau = 1$ or $\tau > 1$ and results related to reducing signal complexity. First, we define terms related to the complexity of STAM systems, and then state our results for signal complexity reduction.

We now provide informal definitions for fan-out and mutual activation. For more rigorous definitions, see [14].

Definition 2. *For an STAM system $\mathcal{T} = (T, \sigma, \tau)$, we say that $\mathcal{T}$ contains* **fan-out** *iff there exists a glue g on a tile $t \in T$ such that whenever g binds, it triggers the activation or deactivation of more than 1 glue on t.*

Definition 3. *For an STAM system* $\mathcal{T} = (T, \sigma, \tau)$, *we say that* $\mathcal{T}$ *contains mutual activation iff* $\exists t_1, t_2 \in T$ *with glue* g *on adjacent edges of* t_1 *and* t_2 *such that whenever* t_1 *and* t_2 *bind by means of glue* g, *the binding of* g *causes the activation or deactivation of other glues on both* t_1 *and* t_2.

3.1 Impossibility of Eliminating Both Fan-Out and Mutual Activation at $\tau = 1$

We now discuss the impossibility of completely eliminating both fan-out and mutual activation at temperature 1. Consider the signal tiles in Figure 1 and let $\mathcal{T} = (T, 1)$ be the STAM$^+$ system where T consists of exactly those tiles. Theorem 1 shows that at temperature 1, it is impossible to completely eliminate both fan-out and mutual activation. In other words, any STAM$^+$ simulation of $\mathcal{T}$ must contain some instance of either fan-out or mutual activation. The intuitive idea is that the only mechanism for turning on glues is binding, and at temperature 1 we cannot control

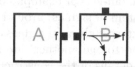

Fig. 1. An example of a tile set where fan-out and mutual activation cannot be completely removed. The glue f on the west edge of tile type B signals two other glues.

when glues in the on state bind. Hence any binding pair of glues that triggers some other glue must do so by means of a sequence of glue bindings leading from the source of the signal to the signal to be turned on. Hence there must be paths to both of the triggered glues from the single originating glue where at some point a single binding event fires two signals. We will see that this is not the case at temperature 2 since we can control glue binding through cooperation there.

Theorem 1. *At temperature 1, there exists an STAM$^+$ system* $\mathcal{T}$ *such that any STAM$^+$ system* $\mathcal{S}$ *that simulates* $\mathcal{T}$ *contains fan-out or mutual activation.*

The proof of Theorem 1 can be found in [14].

3.2 Eliminating Either Fan-Out or Mutual Activation

In this section we will discuss the possibility of eliminating fan-out from an STAM$^+$ system. We do this by simulating a given STAM$^+$ system with a simplified STAM$^+$ system that contains no fan-out, but does contain mutual activation. A slight modification to the construction that we provide then shows that mutual activation can be swapped for fan-out.

Definition 4. *An* n-*simplified STAM tile set is an STAM tile set which has the following properties: (1) the full-tile signal complexity is limited to a fixed constant* $n \in \mathbb{N}$, (2) *there is no fan-out, and (3) fan-in is limited to 2. We say that an STAM system* $\mathcal{T} = (T, \sigma, \tau)$ *is* n-*simplified if* T *is* n-*simplified.*

Theorem 2. *For every STAM$^+$ system $\mathcal{T} = (T, \sigma, \tau)$, there exists a 2-simplified STAM$^+$ system $\mathcal{S} = (S, \sigma', \tau)$ which simulates $\mathcal{T}$ with scale factor $O(|T|^2)$ and tile complexity $O(|T|^2)$.*

To prove Theorem 2, we construct a macrotile such that every pair of signal paths that run in parallel are never contained on the same tile. This means that at most two signals are ever on one tile since it is possible for a tile to contain at most two non-parallel (i.e. crossing) signals. In place of fan-out, we use mutual activation gadgets (see Figure 3) within the *fan-out zone*. Similarly, we use a *fan-in zone* consisting of tiles that merge incoming signals two at a time, in order to reduce fan-in. For examples of these zones, see Figure 2. Next, we print a circuit (a system of signals) around the perimeter of the macrotile which ensures that the external glues (the glues on the edges of the macrotiles that cause macrotiles to bind to one another) are not turned on until a macrotile is fully assembled. More details of the construction can be found in [14] .

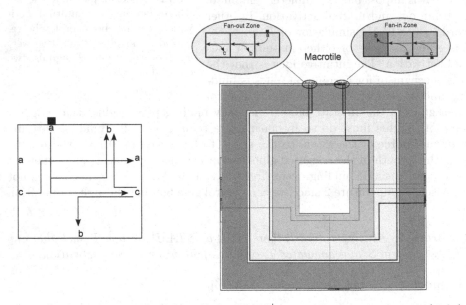

Fig. 2. A tile with 5 signals (left) and the STAM$^+$ macrotile that simulates it (right). Here, the yellow squares represent glue a, the blue square represents glue b and the orange squares represent glue c. The color of each frame corresponds to the glue of the same color. For example on the tile to be simulated (left) there is a signal that runs from glue a to glue c. In order to simulate this signaling, a signal runs from the fan-out zone of glue a (the yellow glue) to the frame associated with glue a on the north edge. The signal then wraps around the frame until it reaches the east side on which glue c lies. Then the signal enters the fan-in zone of glue c.

To further minimize the number of signals per tile at $\tau > 1$, cooperation allows us to reduce the number of signals per tile required to just 1. To achieve this result, we modify the construction used to show Theorem 2, and prove Theorem 3. The details of the modification are in [14] .

Theorem 3. *For every STAM$^+$ system $\mathcal{T} = (T, \sigma, \tau)$ with $\tau > 1$, there exists a 1-simplified STAM$^+$ system $\mathcal{S} = (S, \sigma', \tau)$ which simulates $\mathcal{T}$ with scale factor $O(|T|^2)$ and tile complexity $O(|T|^2)$.*

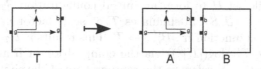

Fig. 3. An example of a mutual activation gadget consisting of tiles A and B without fan-out simulating, at $\tau = 1$, the functionality of tile T which has fan-out. The glue b represents the generic glues which holds the macrotile together. The idea is to "split" the signals from the west glue g on tile A into two signals without using fan-out. Once the west glue g on tile A binds, it turns on the east glue g on tile A. Then, when the east glue g on tile A binds to tile B, it triggers glue f. Thus, the east glue g triggers both the west glue g and glue f without fan-out.

3.3 Summary of Results

At temperature 1, the minimum signal complexity obtainable in general is 2 and while it is possible to eliminate either fan-in or mutual activation, it is impossible to eliminate both. For temperatures greater than 1, cooperation allows for signal complexity to be reduced to just 1 and for both fan-in and mutual activation to be completely eliminated. Table 1 gives a summary of these two cases of reducing signal complexity and shows the cost of such reductions in terms of scale factor and tile complexity.

Table 1. The cost of reducing signal complexity at $\tau = 1$ and at $\tau > 1$

Temperature	Signal per Tile	Scale Factor	Tile Complexity	Contains Fan-In / Mutual Activation
1	2	$O(|T|^2)$	$O(|T|^2)$	one or the other
> 1	1	$O(|T|^2)$	$O(|T|^2)$	neither

4 A 3D 2HAM Tile Set Which Is IU for the STAM$^+$

In this section we present our main result, namely a 3D 2HAM tile set which can be configured to simulate any temperature 1 or 2 STAM$^+$ system, at temperature 2. It is notable that although three dimensions are fundamentally required by the simulations, only two planes of the third dimension are required.

Theorem 4. *There is a 3D tile set U such that, in the 2HAM, U is intrinsically universal at temperature 2 for the class of all 2D STAM$^+$ systems where $\tau \in \{1, 2\}$. Further, U uses no more than 2 planes of the third dimension.*

To prove Theorem 4, we let $\mathcal{T}' = (T', S', \tau)$ be an arbitrary STAM$^+$ system where $\tau \in \{1, 2\}$. For the first step of our simulation, we define $\mathcal{T} = (T, S, \tau)$ as a 2-simplified STAM$^+$ system which simulates $\mathcal{T}'$ at scale factor $m' = O(|T'|^2)$, tile complexity $O(|T'|^2)$, as given by Theorem 2, and let the representation function for that simulation be $R' : B_{m'}^T \dashrightarrow T'$. We now show how to use tiles from a single, universal tile set U to form an initial configuration S_T so that the 3D 2HAM system $\mathcal{U}_T = (U, S_T, 2)$ simulates $\mathcal{T}$ at scale factor $m = O(|T| \log |T|)$ under representation function $R : B_m^U \dashrightarrow T$. This results in $\mathcal{U}_T$ simulating $\mathcal{T}'$ at a scale factor of $O(|T'|^4 \log(|T'|^2))$ via the composition of R and R'. Note that throughout this section, τ refers to the temperature of the simulated systems $\mathcal{T}$ and $\mathcal{T}'$, while the temperature of $\mathcal{U}_{T'}$ is always 2.

4.1   Construction Overview

In this section, due to restricted space we present the 3D 2HAM construction at a very high level. Please see [14] for more details.

Assuming that T is a 2-simplified STAM$^+$ tile set derived from T', we note that for each tile in T: 1. glue deactivation is not used, 2. it has ≤ 2 signals, 3. it has no fan-out, and 4. fan-in is limited to 2. To simulate $\mathcal{T}$, we create an input supertile σ_T from tiles in U so that σ_T fully encodes $\mathcal{T}$ in a rectangular assembly where each row fully encodes the definition of a single tile type from T. Beginning with an initial configuration containing an infinite count of that supertile and the individual tile types from U, assembly begins with the growth of a row on top of (i.e. in the $z = 1$ plane) each copy of σ_T. The tiles forming this row nondeterministically select a tile type $t \in T$ for the growing supertile to simulate, allowing each supertile the possibility of simulating exactly one $t \in T$, and each such t to be simulated. Once enough tiles have attached, that supertile maps to the selected t via the representation function R, and at this point we call it a *macrotile*.

Each such macrotile grows as an extension of σ_T in $z = 0$ to form a square ring with a hole in the center. The growth occurs clockwise from σ_T, creating the west, north, east, then south sides, in that order. As each side grows, the information from the definition of t which is relevant to that side is rotated so that it is presented on the exterior edge of the forming macrotile. The second to last stage of growth for each side is the growth of geometric "bumps and dents" near the corners, which ensure that any two macrotiles which attempt to combine along adjacent edges must have their edges in perfect alignment for any binding to occur. The final stage of growth for each side is to place the glues which face the exterior of the macrotile and are positioned correctly to represent the glues which begin in the on state for that side.

Once the first side of a macrotile completes (which is most likely to be the west side, but due to the nondeterministic ordering of tile additions it could potentially be any side), that macrotile can potentially bind to another macrotile, as long as the tiles that they represent would have been able to do so in $\mathcal{T}$. Whenever macrotiles do bind to each other, the points at which any binding glues exist allow for the attachment of duples (supertiles consisting of exactly 2

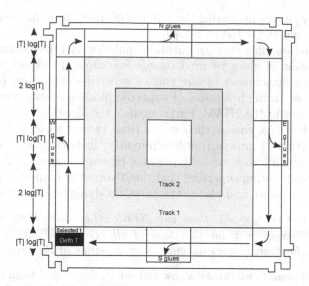

Fig. 4. A high level sketch of the components and formation of a macrotile, including dimensions, not represented to scale

tiles) on top of the two binding tiles (in $z = 1$). These duples initiate the growth of rows in $z = 1$ which move inward on each macrotile to determine if there is information encoded which specifies a signal for that simulated glue to fire. If not, that row terminates. If so, it continues growth by reading the information about that signal (i.e. the destination side and glue), and then growth continues which carries that information inward to the hole in the center of the macrotile. Once there, it grows clockwise in $z = 0$ until arriving at the correct side and glue, where it proceeds to initiate the growth of a row in $z = 1$ out to the edge of the macrotile in the position representing the correct glue. Once it arrives, it initiates the addition of tiles which effectively change the state of the glue from latent to on by exposing the necessary glue(s) to the exterior of the macrotile.

The width of the center hole is carefully specified to allow for the maximum necessary 2 "tracks" along which fired signals can travel, and growth of the signal paths is carefully designed to occur in a zig-zag pattern such that there are well-defined "points of competition" which allow two signals which are possibly using the same track to avoid collisions, with the second signal to arrive growing over the first, rotating toward the next inward track, and then continuing along that track. Further, the positioning of the areas representing the glues on each edge is such that there is always guaranteed to be enough room for the signals to perform the necessary rotations, inward, and outward growth. If it is the case that both signals are attempting to activate the same glue on the same side, when the second signal arrives, the row growing from the innermost track toward the edge of the macrotile will simply run into the "activating" row from the first signal and halt, since there is no need for both to arrive and in the STAM such a situation simply entails that signal being discarded. (Note that this construction can be modified to allow for any arbitrary full-tile signal complexity n for a given tile set by simply increasing the number of tracks to n, and all growth will remain correct and restricted to $z \in \{0, 1\}$.)

This construction allows for the faithful simulation of $\mathcal{T}$ by exploiting the fact that the activation of glues by fired signals is completely asynchronous in the STAM, as is the attachment of any pair of supertiles, and both processes are being represented through a series of supertile binding events which are similarly asynchronous in the 2HAM. Further, since the signals of the STAM$^+$ only ever activate glues (i.e. change their states from latent to on), the constantly "forward" direction of growth (until terminality) in both models ensures that the simulation by $\mathcal{U}_{\mathcal{T}}$ can eventually produce representations of all supertiles in $\mathcal{T}$, while never generating supertiles that don't correctly map to supertiles in $\mathcal{T}$ (equivalent production), and also that equivalent dynamics are preserved.

Theorem 5. *For every $\tau > 1$, there is a 3D tile set U_τ such that, in the 2HAM, U_τ is IU at temperature τ for the class of all 2D STAM$^+$ systems where of temperature τ. Further, U uses no more than 2 planes of the third dimension.*

To prove Theorem 5, we create a new tile set U_τ for each τ from the tile set of Theorem 4 by simply creating $O(\tau)$ new tile types which can encode the value of the strength of the glues of T in $\sigma_{\mathcal{T}}$, and which can also be used to propagate that information to the edges of the macrotiles. For the exterior glues of the macrotiles, just as strength 2 glues were split across two tiles on the exterior of the macrotiles, so will τ-strength glues, with one being of strength $\lceil \tau/2 \rceil$ and the other $\lfloor \tau/2 \rfloor$. All glues which appear on the interior of the macrotile are changed so that, if they were strength 1 glues they become strength $\lceil \tau/2 \rceil$, and if they were strength 2 they become strength τ. In this way, the new tile set U_τ will form macrotiles exactly as before, while correctly encoding glues of strengths 1 through τ on their exteriors, and the systems using it will correctly simulate STAM$^+$ systems at temperature τ.

5   Conclusion

We have shown how to transform STAM$^+$ systems (at temperature 1 or > 1) of arbitrary signal complexity into STAM$^+$ systems which simulate them while having signal complexity no greater than 2 and 1, respectively. However, if the original tile set being simulated is T, the scale factor and tile complexity of the simulating system are approximately $O(|T|^2)$. It seems that these factors cannot be reduced in the worst case, i.e. when a tile of T has a copy of every glue of the tile set on each side, and each copy of each glue on the tile activates every other, yielding a signal complexity of $O(|T|^2)$. However, whether or not this is a true lower bound remains open, as well as what factors can be achieved for more "typical" systems with much lower signal complexity.

A significant open problem which remains is that of generalizing both constructions (the signal reduction and the 3D 2HAM simulation) to the unrestricted STAM. Essentially, this means correctly handling glue deactivation and possible subassembly dissociation. While this can't be handled within the standard 3D 2HAM where glue bonds never change or break, it could perhaps be possible if negative strength (i.e. repulsive) glues are allowed (see [10] for a discussion of various formulations of models with negative strength glues). However, it appears that since both constructions use scaled up macrotiles to represent individual tiles of the systems being simulated, there is a fundamental barrier. The STAM assumes that whenever two tiles are adjacent, all pairs of matching glues across the adjacent edge which are both currently on will immediately bind (which is in contrast to other aspects of the model, which are asynchronous). Since both constructions trade the ability of individual tile edges in the STAM to have multiple glues with scaled up macrotiles which distribute those glues across individual tiles of the macrotile edges, it appears to be difficult if not impossible to maintain the correct simulation dynamics. Basically, a partially formed side of a macrotile could have only a subset of its initially on glues in place, but enough to allow it to bind to another macrotile. At that point, if glue deactivations are initiated which result in the dissociation of the macrotile before the remaining glues of the incomplete macrotile side assemble, then in the simulating system, those additional glues won't ever bind. However, in the simulated system they would have. This results in a situation where, after the dissociation, the simulated system would potentially have additional pending glue actions (initiated by the bindings of the additional glues) which the simulating system would not, breaking the simulation.

Overall, laboratory experiments continue to show the plausibility of physically implementing signalling tiles [17], while previous theoretical work [18] shows some of their potential, and the results in this paper demonstrate how to obtain much of that power with simplified tiles. We feel that research into self-assembly with active components has a huge amount of potential for future development, and continued studies into the various tradeoffs (i.e. complexity of components, number of unique component types, scale factor, etc.) between related models provide important context for such research. We hope that our results help to contribute to continued advances in both theoretical and experimental work along these lines.

References

1. Abel, Z., Benbernou, N., Damian, M., Demaine, E.D., Demaine, M.L., Flatland, R., Kominers, S.D., Schweller, R.T.: Shape replication through self-assembly and rnase enzymes. In: Proceedings of the Twenty-First Annual ACM-SIAM Symposium on Discrete Algorithms, pp. 1045–1064 (2010)
2. Becker, F., Rapaport, I., Rémila, É.: Self-assemblying classes of shapes with a minimum number of tiles, and in optimal time. In: Arun-Kumar, S., Garg, N. (eds.) FSTTCS 2006. LNCS, vol. 4337, pp. 45–56. Springer, Heidelberg (2006)
3. Chandran, H., Gopalkrishnan, N., Reif, J.H.: The tile complexity of linear assemblies. In: Albers, S., Marchetti-Spaccamela, A., Matias, Y., Nikoletseas, S., Thomas, W. (eds.) ICALP 2009, Part I. LNCS, vol. 5555, pp. 235–253. Springer, Heidelberg (2009)
4. Cheng, Q., Aggarwal, G., Goldwasser, M.H., Kao, M.-Y., Schweller, R.T., de Espanés, P.M.: Complexities for generalized models of self-assembly. SIAM Journal on Computing 34, 1493–1515 (2005)
5. Costa Santini, C., Bath, J., Tyrrell, A.M., Turberfield, A.J.: A clocked finite state machine built from DNA. Chem. Commun. 49, 237–239 (2013)

6. Delorme, M., Mazoyer, J., Ollinger, N., Theyssier, G.: Bulking i: An abstract theory of bulking. Theor. Comput. Sci. 412(30), 3866–3880 (2011)
7. Delorme, M., Mazoyer, J., Ollinger, N., Theyssier, G.: Bulking ii: Classifications of cellular automata. Theor. Comput. Sci. 412(30), 3881–3905 (2011)
8. Demaine, E.D., Demaine, M.L., Fekete, S.P., Ishaque, M., Rafalin, E., Schweller, R.T., Souvaine, D.L.: Staged self-assembly: nanomanufacture of arbitrary shapes with $O(1)$ glues. Natural Computing 7(3), 347–370 (2008)
9. Demaine, E.D., Patitz, M.J., Rogers, T.A., Schweller, R.T., Summers, S.M., Woods, D.: The two-handed tile assembly model is not intrinsically universal. In: Fomin, F.V., Freivalds, R., Kwiatkowska, M., Peleg, D. (eds.) ICALP 2013, Part I. LNCS, vol. 7965, pp. 400–412. Springer, Heidelberg (2013)
10. Doty, D., Kari, L., Masson, B.: Negative interactions in irreversible self-assembly. In: Sakakibara, Y., Mi, Y. (eds.) DNA 16. LNCS, vol. 6518, pp. 37–48. Springer, Heidelberg (2011)
11. Doty, D., Lutz, J.H., Patitz, M.J., Schweller, R.T., Summers, S.M., Woods, D.: The tile assembly model is intrinsically universal. In: Proceedings of the 53rd Annual IEEE Symposium on Foundations of Computer Science, FOCS 2012, pp. 302–310 (2012)
12. Fu, B., Patitz, M.J., Schweller, R.T., Sheline, R.: Self-assembly with geometric tiles. In: Czumaj, A., Mehlhorn, K., Pitts, A., Wattenhofer, R. (eds.) ICALP 2012, Part I. LNCS, vol. 7391, pp. 714–725. Springer, Heidelberg (2012)
13. Han, D., Pal, S., Yang, Y., Jiang, S., Nangreave, J., Liu, Y., Yan, H.: DNA gridiron nanostructures based on four-arm junctions. Science 339(6126), 1412–1415 (2013)
14. Hendricks, J.G., Padilla, J.E., Patitz, M.J., Rogers, T.A.: Signal transmission across tile assemblies: 3D static tiles simulate active self-assembly by 2D signal-passing tiles. Technical Report 1306.5005, Computing Research Repository (2013)
15. Ke, Y., Ong, L.L., Shih, W.M., Yin, P.: Three-dimensional structures self-assembled from dna bricks. Science 338(6111), 1177–1183 (2012)
16. Kim, J.-W., Kim, J.-H., Deaton, R.: DNA-linked nanoparticle building blocks for programmable matter. Angewandte Chemie International Edition 50(39), 9185–9190 (2011)
17. Padilla, J.E.: Personal communication (2013)
18. Padilla, J.E., Patitz, M.J., Pena, R., Schweller, R.T., Seeman, N.C., Sheline, R., Summers, S.M., Zhong, X.: Asynchronous signal passing for tile self-assembly: Fuel efficient computation and efficient assembly of shapes. In: Mauri, G., Dennunzio, A., Manzoni, L., Porreca, A.E. (eds.) UCNC 2013. LNCS, vol. 7956, pp. 174–185. Springer, Heidelberg (2013)
19. Pinheiro, A.V., Han, D., Shih, W.M., Yan, H.: Challenges and opportunities for structural dna nanotechnology. Nature Nanotechnology 6(12), 763–772 (2011)
20. Rothemund, P.W., Papadakis, N., Winfree, E.: Algorithmic self-assembly of dna sierpinski triangles. PLoS Biology 2(12), e424 (2004)
21. Winfree, E.: Algorithmic Self-Assembly of DNA. PhD thesis, California Institute of Technology (June 1998)
22. Woods, D., Chen, H.-L., Goodfriend, S., Dabby, N., Winfree, E., Yin, P.: Active self-assembly of algorithmic shapes and patterns in polylogarithmic time. In: Proceedings of the 4th Conference on Innovations in Theoretical Computer Science, ITCS 2013, pp. 353–354. ACM, New York (2013)

3-Color Bounded Patterned Self-assembly*

(Extended Abstract)

Lila Kari[1], Steffen Kopecki[1], and Shinnosuke Seki[2]

[1] Department of Computer Science, The University of Western Ontario,
London, Ontario N6A 5B7, Canada
lila@csd.uwo.ca, steffen@csd.uwo.ca
[2] Helsinki Institute of Information Technology (HIIT),
Department of Computer Science, Aalto University,
P.O. Box 15400, FI-00076, Aalto, Finland
shinnosuke.seki@aalto.fi

Abstract. Patterned self-assembly tile set synthesis (PATS) is the problem of finding a minimal tile set which uniquely self-assembles into a given pattern. Czeizler and Popa proved the NP-completeness of PATS and Seki showed that the PATS problem is already NP-complete for patterns with 60 colors. In search for the minimal number of colors such that PATS remains NP-complete, we introduce multiple bound PATS (MBPATS) where we allow bounds for the numbers of tile types of each color. We show that MBPATS is NP-complete for patterns with just three colors and, as a byproduct of this result, we also obtain a novel proof for the NP-completeness of PATS which is more concise than the previous proofs.

1 Introduction

Tile self-assembly is the autonomous formation of a structure from individual *tiles* controlled by local attachment rules. One application of self-assembly is the implementation of nanoscopic tiles by DNA strands forming double crossover tiles with four unbounded single strands [10]. The unbounded single strands control the assembly of the structure as two, or more, tiles can attach to each other only if the bonding strength between these single strands is big enough. The general concept is to have many copies of the same tile types in a solution which then form a large crystal-like structure over time; often an initial structure, the *seed*, is present in the solution from which the assembly process starts.

A mathematical model describing self-assembly systems is the *abstract tile self-assembly model* (aTAM), introduced by Winfree [9]. Many variants of aTAMs have been studied: a main distinction between the variants is whether the *shape* or the *pattern* of a self-assembled structure is studied. In this paper we focus on the self-assembly of patterns, where a property, modeled as color, is assigned to

* The research of L. K. and S. K. was supported by the NSERC Discovery Grant R2824A01 and UWO Faculty of Science grant to L. K. The research of S. S. was supported by the HIIT Pump Priming Project Grant 902184/T30606.

D. Soloveichik and B. Yurke (Eds.): DNA 2013, LNCS 8141, pp. 105–117, 2013.

each tile; see for example [6] where fluorescently labeled DNA tiles self-assemble into Sierpinski triangles. Formally, a pattern is a rectilinear grid where each vertex has a color: a k-colored $m \times n$-pattern P can be seen as a function $P \colon [m] \times [n] \to [k]$, where $[i] = \{1, 2, \ldots, i\}$. The optimization problem of *patterned self-assembly tile set synthesis* (PATS), introduced by Ma and Lombardi [4], is to determine the minimal number of tile types needed to uniquely self-assemble a given pattern starting from an L-shaped seed. In this paper, we consider the decision variant of PATS, defined as follows:

Problem. (k-PATS)
  GIVEN: A k-colored pattern P and an integer m;
  OUTPUT: "Yes" if P can uniquely be self-assembled by using m tile types.

Czeizler and Popa proved that PATS, where the number of colors on an input pattern is not bounded, is NP-hard [1], but the practical interest lies in k-PATS. Seki proved 60-PATS is NP-hard [8]. By the nature of the biological implementations, the number of distinct colors in a pattern can be considered small. In search for the minimal number k for which k-PATS remains NP-hard, we investigate a modification of PATS: *multiple bound* PATS (MBPATS) uses individual bounds for the number of tile types of each color.

Problem. (k-MBPATS)
  GIVEN: A pattern P with colors from $[k]$ and $m_1, \ldots, m_k \in \mathbb{N}$;
  OUTPUT: "Yes" if P can uniquely be self-assembled by using m_i tile types of color i, for $i \in [k]$.

The main contribution of this paper is a polynomial-time reduction from PATS to 3-MBPATS which proves the NP-hardness of 3-MBPATS. However, our reduction does not take every pattern as input, we only consider a restricted subset of patterns for which PATS is known to remain NP-hard. The patterns we use as input are exactly those patterns that are generated by a polynomial-time reduction from 3-SAT to PATS. Using one of the reductions which were presented in [1,8] as a foundation for our main result turned out to be unfeasible. Therefore, we present a novel proof for the NP-hardness of PATS which serves well as foundation for our main result. Furthermore, our reduction from 3-SAT to PATS is more concise compared to previous reductions in the sense that in order to self-assemble a pattern P we only allow three more tile types than colors in P. In Czeizler and Popa's approach the number of additional tile types is linear in the size of the input formula and Seki uses 84 tile types with 60 colors.

Let us note first that the decision variants of PATS and MBPATS can be solved in NP by simple "guess and check" algorithms. Before we prove NP-hardness of k-PATS, in Sect. 3, and 3-MBPATS, in Sect. 4, we introduce the formal concepts of patterned tile assembly systems, in Sect. 2. We only present some shortened proofs for our lemmas. Full proofs for all lemmas as well as additional figures, depicting our patter designs, can be found in the arXiv version [3].

2   Rectilinear Tile Assembly Systems

In this section we formally introduce patterns and rectilinear tile assembly systems. An excellent introduction to the fundamental model aTAM is given in [7].

Let C be a finite *alphabet of colors*. An $m \times n$-*pattern* P, for $m, n \in \mathbb{N}$, with colors from C is a mapping $P : [m] \times [n] \to C$. By $C(P) \subseteq C$ we denote the colors in the pattern P, i.e., the codomain or range of the function P. The pattern P is called k-*colored* if $|C(P)| \leq k$. The width and height of P are denoted by $w(P) = m$ and $h(P) = n$, respectively. The pattern is arranged such that position $(1, 1)$ is on the bottom left and position $(m, 1)$ is on the bottom right.

Let Σ be a finite *alphabet of glues*. A *colored Wang tile*, or simply *tile*, $t \in C \times \Sigma^4$ is a unit square with a color from C and four glues from Σ, one on each of its edges. $\chi(t) \in C$ denotes the color of t and $t(N)$, $t(E)$, $t(W)$, and $t(S)$ denote the glues on the north, east, west, and south edges of t, respectively. We also call the south and west glues the *inputs* of t while the north and east glues are called *outputs* of t.

A *rectilinear tile assembly system* (RTAS) (T, σ) over C and Σ consists of a set of colored Wang tiles $T \subseteq C \times \Sigma^4$ and an L-shaped seed σ. The seed σ covers positions $(0, 0)$ to $(m, 0)$ and $(0, 1)$ to $(0, n)$ of a two-dimensional Cartesian grid and it has north glues from Σ on the positions $(1, 0)$ to $(m, 0)$ and east glues from Σ on positions $(0, 1)$ to $(0, n)$. We will frequently call T an RTAS without explicitly mentioning the seed. The RTAS T describes the self-assembly of a structure: starting with the seed, a tile t from T can attach to the structure at position $(x, y) \in [m] \times [n]$, if its west neighbor at position $(x - 1, y)$ and south neighbor at position $(x, y - 1)$ are present and the inputs of t match the adjacent outputs of its south and west neighbors; the self-assembly stops when no more tiles in T can be attached by this rule. Arbitrarily many copies of a each tile type in T are considered to be present while the structure is self-assembled, thus, one tile type can appear in multiple positions. A *tile assignment* in T is a function $f : [m] \times [n] \to T$ such that $f(x, y)(W) = f(x - 1, y)(E)$ and $f(x, y)(S) = f(x, y - 1)(N)$ for $(x, y) \in [m] \times [n]$. The RTAS self-assembles a pattern P if there is a tile assignment f in T such that the color of each tile in the assignment f is the color of the corresponding position in P, i.e., $\chi \circ f = P$. A terminological convention is to call the elements in T *tile types* while the elements in a tile assignment are called *tiles*.

A *directed RTAS* (DRTAS) T is an RTAS where any two distinct tile types $t_1, t_2 \in T$ have different inputs, i.e., $t_1(S) \neq t_2(S)$ or $t_1(W) \neq t_2(W)$. A DRTAS has at most one tile assignment and can self-assemble at most one pattern. If T self-assembles an $m \times n$-pattern P, it defines the function $P_T : [m] \times [n] \to T$ such that $P_T(x, y)$ is the tile in position (x, y) of the tile assignment given by T. In this paper, we investigate minimal RTASs which uniquely self-assemble one given pattern P. As observed in [2], if P can be uniquely self-assembled by an RTAS with m tile types, then P can also be (uniquely) self-assembled by a DRTAS with m tile types.

3 NP-Hardness of Pats

In this section, we prove the NP-hardness of Pats. The proof we present uses many techniques that have already been employed in [1,8]. Let us also point out that we do not intend to minimize the number of colors used in our patterns or the size of our patterns. Our motivation is to give a proof that is easy to understand and serves well as a foundation for the results in Sect. 4.

A boolean formula F over variables V in *conjunctive normal form with three literals per clause*, 3-CNF for short, is a boolean formula such that

$$F = (c_{1,1} \lor c_{1,2} \lor c_{1,3}) \land (c_{2,1} \lor c_{2,2} \lor c_{2,3}) \land \cdots \land (c_{\ell,1} \lor c_{\ell,2} \lor c_{\ell,3})$$

where $c_{i,j} \in \{v, \neg v \mid v \in V\}$ for $i \in [\ell]$ and $j = 1, 2, 3$. It is well known that the problem 3-Sat, to decide whether or not a given formula F in 3-CNF is satisfiable, is NP-complete; see e.g., [5]. The NP-hardness of Pats follows by the polynomial-time reduction from 3-Sat to Pats, stated in Theorem 1.

Theorem 1. *For every formula F in 3-CNF there exists a pattern P_F such that F is satisfiable if and only if P_F can be self-assembled by a DRTAS with at most $|C(P_F)| + 3$ tile types. Moreover, P_F can be computed from F in polynomial time.*

Theorem 1 follows by Lemmas 3 and 5, which are presented in the following.

The pattern P_F consists of several rectangular *subpatterns* which we will describe in the following. None of the subpatterns will be adjacent to another subpattern. The remainder of the pattern P_F is filled with *unique colors*; a color c is unique in a pattern P if it appears only in one position in P, i.e., $|P^{-1}(c)| = 1$. As a technicality that will become useful only in the proof of Theorem 2, we require that each position adjacent to the L-shaped seed or to the north or east border of pattern P_F has a unique color. Clearly, for each unique color in P_F we require exactly one tile in any DRTAS which self-assembles P_F. Since each subpattern is surrounded by a frame of unique colors, the subpatterns can be treated as if each of them would be adjacent to an L-shaped seed and we do not have to care about the glues on the north border or east border of a subpattern.

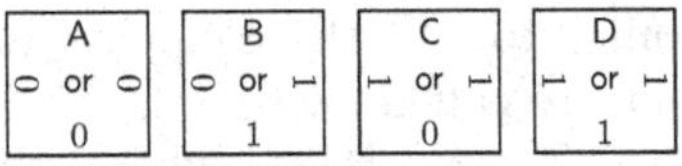

A	B	C	D
○ or ○	○ or ⊢	⊢ or ⊢	⊢ or ⊢
0	1	0	1

Fig. 1. The four tile types used to implement the or-gate

As stated earlier, the number of tile types m that is required to self-assemble P_F, if F is satisfiable, is $m = |C(P_F)| + 3$. Actually, every color in $C(P_F)$ will require one tile type only except for one color which is meant to implement an or-gate; see Fig. 1. Each of the tile types with color $\boxed{\text{or}}$ is supposed to have west input $w \in \{0, 1\}$, south input $s \in \{0, 1\}$, east output $w \lor s$, and an independent north output.

Our first subpattern p, shown in Fig. 2, ensures that every DRTAS which self-assembles the subpattern p contains at least three tile types with color $\boxed{\text{or}}$.

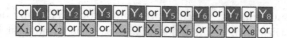

Fig. 2. The subpattern p

For the upcoming proof of Theorem 2 we need a more precise observation which draws a connection between the number of distinct output glues and the number of distinct tile types with color [or].

Lemma 1. *A DRTAS T which self-assembles a pattern including the subpattern p contains either*

- *i.) three distinct tile types $o_1, o_2, o_3 \in T$ with color [or] all having distinct north and east glues,*
- *ii.) four distinct tile types $o_1, o_2, o_3, o_4 \in T$ with color [or] all having distinct north glues and together having at least two distinct east glues,*
- *iii.) four distinct tile types $o_1, o_2, o_3, o_4 \in T$ with color [or] all having distinct east glues and together having at least two north glues, or*
- *iv.) eight distinct tile types $o_1, \ldots, o_8 \in T$ with color [or] all having distinct east or north glues.*

Lemma 1 follows by the fact that each of the tiles with colors [Y1] to [Y8] has the or-gate as west and south neighbors, hence, the number of east glues times the number of north glues of all tile types with color [or] has to be at least eight.

We aim to have statement *ii.)* of Lemma 1 satisfied, but so far all four statements are possible. The subpatterns q_1 to q_5 in Fig. 3 will enforce the functionality of the or-gate tile types.

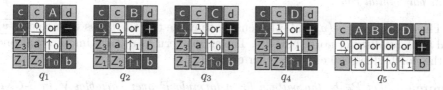

Fig. 3. The subpatterns q_1 to q_5

Lemma 2. *Let P be a pattern that contains the subpatterns p and q_1 to q_5, and let $m = |C(P)| + 3$. A DRTAS T with at most m tile types which self-assembles pattern P contains four tile types with color [or] of the forms shown in Fig. 1. For every other color in $C(P)$ there exists exactly one tile type in T. Moreover, the tile type with color [⁰→] has east output 0 and the tile type with color [+] has west input 1.*

There are at least three or-gate tile types, thus, only the color of one tile type in T is not determined yet. The clue of patterns q_1 to q_4 is that if, e. g., the two tiles with colors [c] in q_1 and q_2 were of different types, there would be only one tile type of the other colors, and in particular, their west neighbors would be of the same type as well as their south neighbors. Thus, these two tile types would have the same inputs, which is prohibited for DRTAS by definition. This implies that the tiles with colors A and B have the same west input and can only be placed because their south neighbors, the or-gate tiles in q_1 and q_2, are of different types. By analogous arguments the four or-gate tiles in q_1 to q_4 are of four different types. Subpattern q_5 ensures that the east and west glues of the or-gates match in the way shown in Fig. 1.

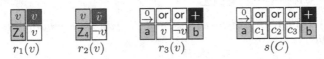

Fig. 4. The subpatterns $r_1(v)$ to $r_3(v)$ for a variable $v \in V$ and the subpattern $s(C)$ for a clause $C = (c_1 \vee c_2 \vee c_3)$ in F where $c_i = v$ or $c_i = \neg v$ for some variable $v \in V$ and $i = 1, 2, 3$.

The subpatterns that we defined so far did not depend on the formula F. Now, for each variable $v \in V$ we define three subpatterns $r_1(v)$, $r_2(v)$, $r_3(v)$ and for a clause C from F we define one more subpattern $s(C)$; these patterns are given by Fig. 4. For a formula F in 3-CNF we let P_F be the pattern that contains all the subpatterns p, q_1 to q_5, $r_1(v)$ to $r_3(v)$ for each variable $v \in V$, and $s(C)$ for each clause C from F, where each subpattern is placed to the right of the previous subpattern with one column of unique colors in between. Then, P_F has height 6, because the top and bottom rows contain unique colors only, and P_F has width $45 + 11 \cdot |V| + 6 \cdot \ell$. The next lemma follows from this observation.

Lemma 3. *Given a formula F in 3-CNF, the pattern P_F can be computed from F in polynomial time.*

The subpatterns $r_1(v)$ and $r_2(v)$ ensure that the two tile types with colors v and $\neg v$ have distinct north outputs. The subpattern $r_3(v)$ then implies that one of the north glues is 0 and the other one is 1.

Lemma 4. *Let P_F be the pattern for a formula F over variables V in 3-CNF and let T be a DRTAS with at most $m = |C(P_F)| + 3$ tile types which self-assembles pattern P_F. For all variables $v \in V$, there is a unique tile type $t_v^{\oplus} \in T$ with color v and a unique tile type $t_v^{\ominus} \in T$ with color $\neg v$ such that either $t_v^{\oplus}(N) = 1$ and $t_v^{\ominus}(N) = 0$ or $t_v^{\oplus}(N) = 0$ and $t_v^{\ominus}(N) = 1$.*

Now, these glues serve as input for the or-gates in the subpatterns $s(C)$. The following lemma concludes the proof of Theorem 1.

Lemma 5. *Let P_F be the pattern for a formula F over variables V in 3-CNF and let $m = |C(P_F)| + 3$. The formula F is satisfiable if and only if P_F can be self-assembled by a DRTAS T with at most m tile types.*

The formula F is satisfiable if and only if there is a variable assignment $f : V \to \{0, 1\}$ which satisfies every clause in F. In order for $s(C)$ with $C = (c_1 \vee c_2 \vee c_3)$ to self-assemble, one of the north glues of the tiles for c_1, c_2, or c_3

has to be 1. Let $t_v^\oplus$ and $t_v^\ominus$ for $v \in V$ as before. Since $t_v^\oplus(N)$ and $t_v^\ominus(N)$ represent opposite truth values, the pattern P can be self-assembled using m tile types if and only if $f(v) = t_v^\oplus(N)$ satisfies every clause in F. How the remaining tile types and glues in T can be chosen is shown in the arXiv version [3].

4   NP-Hardness of 3-MBPATS

The purpose of this section is to prove the NP-hardness of 3-MBPATS. Let us define a set of restricted input pairs $\mathcal{I}$ for PATS. The set $\mathcal{I}$ contains all pairs (P, m) where $P = P_F$ is the pattern for a formula F in 3-CNF as defined in Sect. 3 and $m = |C(P)| + 3$. Consider the following restriction of PATS.

Problem. (MODIFIED PATS)
  GIVEN: A pair (P, m) from $\mathcal{I}$;
  OUTPUT: "Yes" if P can uniquely be self-assembled by using m tile types.

As we choose exactly those pairs (P, m) as input for the problem that are generated by the reduction, stated in Theorem 1, we obtain the following corollary which forms the foundation for the result in this section.

Corollary 1. MODIFIED PATS *is NP-hard.*

The NP-hardness of 3-MBPATS follows by the polynomial-time reduction from MODIFIED PATS to 3-MBPATS, stated in Theorem 2.

Theorem 2. *For every input pair $(P, m) \in \mathcal{I}$ there exist a black/white/gray-colored pattern Q and integers m_b, m_w, m_g such that: P can be self-assembled by a DRTAS with at most m tile types if and only if Q can be self-assembled by a DRTAS with at most m_b black tile types, m_w white tile types, and m_g gray tile types. Moreover, the tuple (Q, m_b, m_w, m_g) can be computed from P in polynomial time.*

Lemma 12 states the "if part" and Lemma 8 states the "only if part" of Theorem 2. Lemma 6 states that (Q, m_b, m_w, m_g) can be computed from P in polynomial time.

For the remainder of this section, let $(P, m) \in \mathcal{I}$ be one fixed pair, let $C = C(P)$ and $k = |C|$. We may assume that $C = [k]$ is a subset of the positive integers. The tile bounds are $m_b = 1$ for black tile types, $m_w = 5k - 3(w(P) + h(P)) + 14$ for white tile types, and $m_g = 2k + 3$ for gray tile types. Note that, due to the pattern design in Sect. 3, $h(P) = 6$ is constant.

Let $\ell = 5k + 8$. For a color $c \in C$, we define an $\ell \times \ell$ square pattern as shown in Fig. 5. We refer to this pattern as well as to its underlying tile assignment as *supertile*. In contrast to the previous section, the positions in the supertile are labeled which does not mean that the colors or the tiles used to self-assemble the pattern are labeled; the colors are black, white, or gray. The horizontal and vertical *color counters* are the c gray positions in the top row, respectively right column, which are succeeded by a white tile in position D_2, respectively D_1.

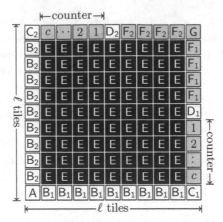

Fig. 5. Black/white/gray supertile which portrays a color $c \in C$

The color counters illustrate the color c that is *portrayed* by the supertile. The patterns of two supertiles which portray two distinct colors differ only in the place the white tile is positioned in its top row and right column.

For colors in the bottom row and left column of the pattern P we use *incomplete supertiles*: a supertile portraying a color c in the bottom row of pattern P lacks the white row with positions A, B_1, and C_1; a supertile representing a color c in the left column of pattern P lacks the white column with positions A, B_2, and C_2. In particular, the supertile portraying color $P(1, 1)$ does not contain any of the positions A, B_1, B_2, C_1, and C_2. Recall that all incomplete supertiles portray a color c that is unique in P.

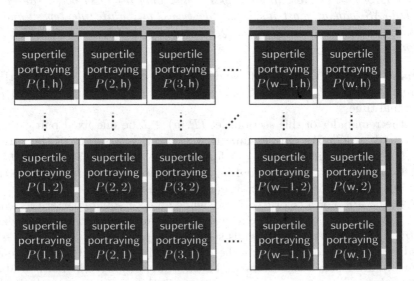

Fig. 6. Black/white/gray pattern Q defined by the k-color pattern P with $\mathsf{w} = w(P)$ and $\mathsf{h} = h(P)$

The pattern Q is shown in Fig. 6. By $Q\langle x, y\rangle$ we denote the pattern of the supertile covering the square area spanned by positions $((x - 1) \cdot \ell, (y - 1) \cdot \ell)$ and $(x \cdot \ell - 1, y \cdot \ell - 1)$ in Q; the incomplete supertiles cover one row and/or column less. The pattern is designed such that supertile $Q\langle x, y\rangle$ portrays the color $P(x, y)$ for all $x \in [w(P)]$ and $y \in [h(P)]$. Additionally, Q contains three *gadget rows* and three *gadget columns* which are explained in Fig. 7. The purpose of these gadget rows and columns is to ensure that the color counters can only be implemented in one way when using no more than m_g gray tile types. All together Q is of dimensions $w(Q) = \ell \cdot w(P) + 2$ times $h(Q) = \ell \cdot h(P) + 2$. Obviously, the pattern Q can be computed from P in polynomial time.

Lemma 6. *(Q, m_b, m_w, m_g) can be computed from P in polynomial time.*

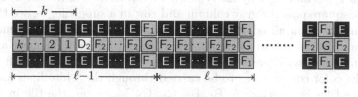

Fig. 7. The gadget rows on the north border of the pattern Q, the gadget columns are symmetrical: the middle row (resp., column) contains gray tiles except for one white tile in position $k + 1$; the upper and lower rows (resp., left and right columns) contain gray tiles in positions above the gray column (resp., right of the gray row) of a supertile, the other tiles are black.

For a DRTAS Θ which self-assembles Q, we extend our previous notion such that $Q_\Theta\langle x, y\rangle$ denotes the tile assignment of supertile $Q\langle x, y\rangle$ given by Θ. In the following, we will prove properties of such a DRTAS Θ. Our first observation is about the black and gray tile types plus two of the white tile types.

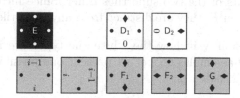

Fig. 8. The black tile type, two of the white tile types, and all gray tile types: the labeled tile types are used in the corresponding positions of each supertile and the gadget pattern; the unlabeled tile types, called *counter tiles* for $i \in [k]$, implement the vertical and horizontal color counters.

Lemma 7. *Let Θ be a DRTAS which self-assembles the pattern Q using at most $m_b = 1$ black tile types and $m_g = 2k + 3$ gray tile types. The black and gray tile types in Θ are of the form shown in Fig. 8 and Θ contains two white tiles of the form shown in the figure. In every supertile, the horizontal and vertical color* *counters are implemented by a subset of the counter tile types and for a position* $E, D_1, D_2, F_1, F_2,$ *or* G *the correspondingly labeled tile type is used. Furthermore, the glues* $\bullet, \blacklozenge, 0, 1, \ldots, k$ *are all distinct.*

Since there is only one black tile type which can tile the black square area in each supertile, the black tile type has to be of the given form. In particular, no kind of information can be passed through the black square areas in the supertiles. The k gray tiles, followed by one white tile in the gadget rows and columns, ensure that some kind of horizontal and vertical counter tile types are present in Θ. The three remaining gray tile types have to be used for positions $F_1, F_2,$ and G; it is easy to see that they are of the given forms.

Remark 1. Consider a DRTAS Θ that self-assembles the pattern Q using most m_b black tile types and m_g gray tile types. If we have a look at the tile assignment of the black square plus the gray column and row in a supertile, we see that this block has inputs $\bullet$ on all edges except for edges where the color counters are initialized and it has outputs $\bullet$ on all edges, except for its right-most and top-most output edges which are $\blacklozenge$. This means that all information on how to initialize the color counters has to be carried through the white lines and rows, that are, the tiles in positions A, B_1, B_2, C_1, C_2. Moreover, the tile in position A is the only one with non-generic input from other supertiles. This tile fully determines the tile assignment of the supertile and can be seen as the *control tile* or *seed* of the supertile. Henceforth, for a supertile $s = Q_\Theta\langle x, y \rangle$ we extend our notion of glues such that $s(S)$ and $s(W)$ denote the south and west input of the tile in position A, respectively, $s(N)$ and $s(E)$ denote the north and east output of the tiles in positions C_2 and C_1, respectively. For incomplete supertiles only one of $s(N)$ or $s(E)$ is defined.

Two supertiles in Q_Θ are considered distinct if their tile assignment differs in at least one position. By the observations above, two complete supertiles are distinct if and only if their control tiles are of distinct types; this is equivalent to require that the inputs of the two supertiles differ. Since incomplete supertiles portray unique colors in P, they are distinct from any supertile in Q_Θ but itself.

There is some flexibility in how the white tile types are implemented in a DRTAS Θ which self-assembles Q. Let us present one possibility which proves the "only if part" of Theorem 2.

Lemma 8. *If* P *can be self-assembled by a DRTAS* T *with* m *tile types, then* Q *can be self-assembled by a DRTAS* Θ *using* m_b *black tile types,* m_w *white tile types, and* m_g *gray tile types.*

Proof. Let Θ contain the tile types given in Fig. 8. For a supertile portraying a color $c \in C \setminus \{\boxed{\text{or}}\}$ we use the five tile types given in Fig. 9. Note that we need less tile types for incomplete supertiles which leads to $5 \cdot (k-1) - 3 \cdot (h(P) + w(P)) + 1$ white tile types in total. Thus, we have 16 white tile types left for the or-gate.

Fig. 9. White tile types for the supertile portraying a color $c \in C$, except for the or-gate, where $t \in T$ with $c = \chi(t)$, $\mathsf{n} = t(N)$, $\mathsf{e} = t(E)$, $\mathsf{s} = t(S)$, and $\mathsf{w} = t(W)$

Since three of the or-gates have the same east output, see Fig. 1, they can share tile types in positions B_1 and C_1. The 16 white tile types in Fig. 10 are used to self-assemble the supertiles representing the or-gates. The tile types are designed such that they can self-assemble pattern Q. □

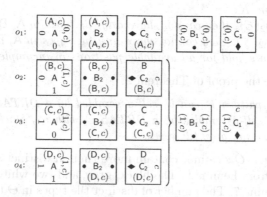

Fig. 10. White tile types for supertiles portraying the or-gate where $o_1, o_2, o_3, o_4 \in T$ are defined in Fig. 1

For the converse implication of Theorem 2, let us show how to obtain a DRTAS that self-assembles P from the supertiles in Q_Θ. The following result follows from the bijection between supertiles in Q_Θ and tiles in P_T.

Lemma 9. *Let Θ be a DRTAS which self-assembles Q using at most m_b black tile types and m_g gray tile types, and let*

$$S = \{Q_\Theta\langle x, y\rangle \mid x \in [w(P)], y \in [h(P)]\}$$

be the set of all distinct supertiles in Q_Θ. There exists a DRTAS T with $|S|$ tile types which self-assembles P such that for each supertile $s \in S$ there exists a tile type $t_s \in T$ with the same glues on the respective edges and s portrays the color of t_s. For an incomplete supertile the statement holds for the defined glue.

We continue with the investigation of the white tile types that are used to self-assemble the pattern Q. The next lemma follows by a case study of what would go wrong if one tile type were used in two of the positions.

Lemma 10. *Let Θ be a DRTAS which self-assembles the pattern Q using at most m_b black tile types and m_g gray tile types. A white tile type from Θ which is used in one of the positions A, B_1, B_2, C_1, C_2, D_1, or D_2 cannot be used in another position in any supertile.*

Let B_1^* be the right-most position B_1 in a supertile, adjacent to position C_1, and let B_2^* be the top-most position B_2 in a supertile, adjacent to position C_2. The following argument is about tiles in the five positions $K = \{A, B_1^*, B_2^*, C_1, C_2\}$ of each supertile. Following Remark 1 it is clear that a tile in position A fully determines the supertile, tiles in positions B_1^* and C_1 carry the color and the east glue of a supertile, whereas tiles in positions B_2^* and C_2 carry the color and the north glue.

Lemma 11. *Let Θ be a DRTAS which self-assembles Q using at most m_b black tile types and m_g gray tile types. Let s_1 and s_2 be supertiles in Q_Θ.*

i.) If s_1 and s_2 portray different colors, they cannot share any tile types in positions from K.

ii.) If $s_1(E) \neq s_2(E)$, they cannot share any tile types in A, B_1^*, or C_1.

iii.) If $s_1(N) \neq s_2(N)$, they cannot share any tile types in A, B_2^*, or C_2.

The three statements hold for all available positions in incomplete supertiles.

Let us conclude the proof of Theorem 2.

Lemma 12. *The pattern P can be self-assembled by a DRTAS T with m tile types if Q can be self-assembled by a DRTAS Θ with m_b black tile types, m_w white tile types, and m_g gray tile types.*

Proof. We show that Q_Θ cannot contain more than m distinct supertiles, then, the claim follows from Lemma 9. The black, gray, and two white tile types in Θ are defined by Lemma 7. The number of distinct tile types in Θ that can be used as control tiles, equals to the number of distinct complete supertiles of Q_Θ. By Lemma 11 we need five white tile types for each complete supertile portraying a color in $C \setminus \{\boxed{\text{or}}\}$; of these five tile types one can be used as control tile. For incomplete supertiles we need just two white tile types, and none for the one supertile portraying $P(1,1)$. There are 16 white tile types left for the or-gate supertiles. From Lemma 1 and Lemma 9 we infer that among these 16 white tile types we can have at most four control tiles. Therefore, the number of distinct supertiles in Q_Θ is $k + 3 = m$ — concluding the proof. □

5   Conclusions

We prove that k-MBPATS, a natural variant of k-PATS, is NP-complete for $k = 3$. Furthermore, we present a novel proof for the NP-completeness of PATS and our proof is more concise than previous proofs. We introduce several new techniques for pattern design in our proofs, in particular in Sect. 4, and we anticipate that these techniques can ultimately be used to prove that 2-MBPATS and also 2-PATS are NP-hard.

References

1. Czeizler, E., Popa, A.: Synthesizing minimal tile sets for complex patterns in the framework of patterned DNA self-assembly. In: Stefanovic, D., Turberfield, A. (eds.) DNA 18. LNCS, vol. 7433, pp. 58–72. Springer, Heidelberg (2012)

2. Göös, M., Orponen, P.: Synthesizing minimal tile sets for patterned DNA self-assembly. In: Sakakibara, Y., Mi, Y. (eds.) DNA 16. LNCS, vol. 6518, pp. 71–82. Springer, Heidelberg (2011)
3. Kari, L., Kopecki, S., Seki, S.: 3-color bounded patterned self-assembly. arXiv preprint arXiv:1306.3257 (2013)
4. Ma, X., Lombardi, F.: Synthesis of tile sets for DNA self-assembly. IEEE Transactions on Computer-Aided Design of Integrated Circuits and Systems 27(5), 963–967 (2008)
5. Papadimitriou, C.H.: Computational complexity. John Wiley and Sons Ltd. (2003)
6. Rothemund, P.W., Papadakis, N., Winfree, E.: Algorithmic self-assembly of DNA Sierpinski triangles. PLoS Biology 2(12), e424 (2004)
7. Rothemund, P.W., Winfree, E.: The program-size complexity of self-assembled squares. In: Proceedings of the Thirty-second Annual ACM Symposium on Theory of Computing, pp. 459–468. ACM (2000)
8. Seki, S.: Combinatorial optimization in pattern assembly. In: Mauri, G., Dennunzio, A., Manzoni, L., Porreca, A.E. (eds.) UCNC 2013. LNCS, vol. 7956, pp. 220–231. Springer, Heidelberg (2013)
9. Winfree, E.: Algorithmic self-assembly of DNA. PhD thesis, California Institute of Technology (1998)
10. Winfree, E., Liu, F., Wenzler, L.A., Seeman, N.C.: Design and self-assembly of two-dimensional DNA crystals. Nature 394(6693), 539–544 (1998)

Exponential Replication of Patterns in the Signal Tile Assembly Model

Alexandra Keenan*, Robert Schweller*, and Xingsi Zhong*

Department of Computer Science
University of Texas - Pan American
{abkeenan,rtschweller,zhongx}@utpa.edu

Abstract. Chemical self-replicators are of considerable interest in the field of nanomanufacturing and as a model for evolution. We introduce the problem of self-replication of rectangular two-dimensional patterns in the practically motivated Signal Tile Assembly Model (STAM) [9]. The STAM is based on the Tile Assembly Model (TAM) which is a mathematical model of self-assembly in which DNA tile monomers may attach to other DNA tile monomers in a programmable way. More abstractly, four-sided tiles are assigned glue types to each edge, and self-assembly occurs when singleton tiles bind to a growing assembly, if the glue types match and the glue binding strength exceeds some threshold. The signal tile extension of the TAM allows signals to be propagated across assemblies to activate glues or break apart assemblies. Here, we construct a pattern replicator that replicates a two-dimensional input pattern over some fixed alphabet of size ϕ with $O(\phi)$ tile types, $O(\phi)$ unique glues, and a signal complexity of $O(1)$. Furthermore, we show that this replication system displays exponential growth in n, the number of replicates of the initial patterned assembly.

1 Introduction

Artificial self-replicating systems have been the subject of various investigations since John von Neumann first outlined a detailed conceptual proposal for a non-biological self-replicating system [7]. Gunter von Kiedrowski, who demonstrated the first enzyme-free abiotic replication system in 1986 [17], describes a model that can be used to conceptualize template-directed self-replication [10]. In this model, minimal template-directed self-replicating systems consist of an auto-catalytic template molecule, and two or more substrate molecules that bind the template molecule and join together to form another template molecule. To date, simple self-replicating systems have been demonstrated in the laboratory with nucleic acids, peptides, and other small organic molecules [11, 16, 17, 19].

Given that substrate molecules must come together without outside guidance to replicate the template, a template-directed self-replicating system is necessarily a self-assembling system. In theoretical computer science, the Tile Assembly

---

* This author's research was supported in part by National Science Foundation Grant CCF-1117672.

D. Soloveichik and B. Yurke (Eds.): DNA 2013, LNCS 8141, pp. 118–132, 2013.

Model (TAM) has become the most commonly used model to describe various self-assembly processes [18]. Many model varients have been described since Erik Winfree first introduced the TAM, however models that are most relevant to self-replicating systems are those that allow for assembly breakage. These include the enzyme staged assembly model [1], the temperature programming model [6], the signal tile assembly model [8,9], and the use of negative glues [12].

Replication of arbitrary 0-genus shapes has been shown within the staged assembly system with the use of RNAse enzymes [1]. Replication and evolution of combinatorial 'genomes' via crystal-like growth and breakage have also been demonstrated in the laboratory using DNA tile monomers [13]. Under this replication mechanism, a DNA crystal ribbon has a sequence of information, or genome, in each row. Upon chance breakage, the daughter crystal continues to grow and copy the genome of the mother crystal. It was further shown that the fidelity of the replication process is sufficiently high for Darwinian evolution. Such simple, enzyme-free systems are of particular importance to the study of the origins of life.

A template-directed method of exponential self-replication within the tile assembly system, where the child molecule detaches from and is identical to the parent (as is found in biological systems), has not yet been described. Here, we present a theoretical basis for template-directed exponential self-replication in the practically motivated Signal Tile Assembly Model (STAM), and in doing so partially address an open question presented by Abel and colleagues [1]. Specifically, we consider the problem of self-replication of rectangular two-dimensional patterns in the STAM. The STAM is a powerful model of tile self-assembly in which activation, via binding, of a glue on an individual tile may turn other glues either on or off elsewhere on the tile [9]. In this way, signals may be propagated across distances greater than a single tile and assemblies may be broken apart. DNA strand displacement reactions provide a plausible physical basis for the signaling cascades used in the STAM. DNA strand displacement occurs when two DNA strands with at least partial complementarity hybridize with each other, which can displace pre-hybridized strands. In the STAM, these reactions may be queued to result in a cascade that ultimately turns a glue "on" by releasing a prehybridized strand. Conversely these queued reactions could turn a glue "off" by binding a free strand, thus making it unavailable to interact with other glues.

An important objective of nanotechnology is to manufacture things inexpensively, thus the prospect of self-replicating materials with useful patterns or functions is enticing. Additionally, an enzyme-free self-replicator that can support and autonomously replicate an information-bearing genome could provide the basis for a model of Darwinian evolution. Because true Darwinian selection necessitates exponential population growth [15], and this rate of growth is also desirable for low-cost manufacturing of nanoscale devices, we approach this problem with the goal of exponential growth in mind.

1.1   Outline of Paper

The Signal Tile Assembly Model of [9] is briefly defined formally in Section 2, followed by our formal definition of exponential replication. In Section 3, we present our main result: there exists a single, general purpose 2D signal tile system that exponentially replicates any rectangular 2D pattern (Theorem 1). We present the signal tile system that achieves this replication, along with a high-level sketch of how the system performs the replication, but omit a detailed analysis due to space limitations in this version.

2   Definitions

2.1   Basic Definitions

Multisets. A multiset is an ordered pair (S, m) where S is a subset of some universe set U and m is a function from U to $\mathbb{N} \bigcup \{\infty\}$ with the property that $m(x) \geq 1$ for all $x \in S$ and $m(x) = 0$ for all $x \notin S$. A multiset models a collection of items in which there are a positive number of copies $m(x)$ of each element x in the collection (called the multiplicity of x). For a multi-set $A = (S, m)$ and $x \in S$, we will use notation $A(x) = m(x)$ to refer to the multiplicity of item x. For multisets $B = (b, m)$ and $A = (a, n)$, define $B \uplus A$ to be the multiset $(a \bigcup b, m')$ where $m'(x) = m(x) + n(x)$. If $m(x) \geq n(x)$ for all $x \in U$, then define $B - A$ to be the multiset $(b', m'(x))$ where $b' = \{x \in b | m(x) - n(x) \geq 1\}$ and $m'(x) = m(x) - n(x)$.

Patterns. Let ϕ be a set of labels that contains at least one particular label null $\in \phi$ which conceptually denotes a blank, non-existent label. Informally, a 2D pattern is defined to be a mapping of 2D coordinates to elements of ϕ. Further, as these patterns will denote patterns on the surface of free floating tile assemblies, we add that patterns are equal up to translation. Formally, a 2D pattern over set ϕ is any set $\{f_{\Delta_x, \Delta_y}(x, y) | \Delta_x, \Delta_y \in \mathbb{Z}\}$ where $f : \mathbb{Z}^2 \to \phi$, and $f_{\Delta_x, \Delta_y}(x, y) = f(x + \Delta_x, y + \Delta_y)$. In this paper we focus on the the class of *rectangular* patterns in which the null label occurs at all positions outside of a rectangular box, with positions within the box labeled arbitrarily with non null labels.

2.2   Signal Tile Model

In this section we define the signal tile assembly model (STAM) by defining the concepts of an *active tile* consisting of a unit square with *glue slots* along the faces of the tile, as well as *assemblies* which consist of a collection of active tiles positioned on the integer lattice. We further define a set of three *reactions* (*break* reactions, *combination* reactions, and *glue-flip* reactions) which define how a set of assemblies can change over time. Figure 1 represents each of these concepts pictorially to help clarify the following technical definitions. Please see [9] for a more detailed presentation of the STAM.

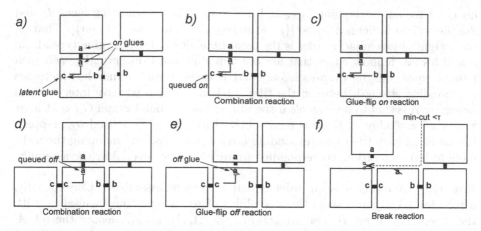

Fig. 1. This sequence (a-f) demonstrates the reaction types, glue states, and queued commands defined in the STAM

Glue Slots. Glue slots are the signal tile equivalent of glues in the standard tile assembly model with the added functionality of being able to be in one of three states, *on, off,* or *latent,* as well as having a *queued command* of *on, off,* or -, denoting if the glue is queued to be turned on, turned off, or has not been queued to change state. Formally, we denote a glue slot as an ordered triple $(g, s, q) \in \Sigma \times \{on, off, latent\} \times \{on, off, -\}$ where Σ is some given set of labels referred to as the *glue type* alphabet. For a given glue slot $x = (g, s, q)$, we define the *type* of x to be g, the *state* of x to be s, and the *queued action* of x to be q.

Active Tiles. An active tile is a 4-sided unit square with each edge having a sequence of *glue slots* $g_1, \ldots g_r$ for some positive integer r, as well as an additional *label* taken from a set of symbols ϕ. For simplicity of the model, we further require that the glue type of each g_i on each tile face is the same (although state and queued commands may be different), and that the glue type of g_i is distinct from the glue type of g_j if $i \neq j$. For an active tile t, let $t_{d,i}$ denote the glue slot g_i on face d of active tile t.

Finally, an active tile t has an associated *signal function* $f_t(d, i)$ which assigns to each glue slot i on each tile side d a corresponding set of triples consisting of a glue slot, a side, and a command, which together denote which glue slots of each tile face should be turned on or off in the effect that slot i on face d becomes bonded. Formally, each active tile t has an associated *signal function* $f : \{north, south, east, west\} \times \{1, \ldots r\} \to \mathcal{P}(\{north, east, south, west\} \times \{1, \ldots r\} \times \{on, off\})$. For the remainder of this paper we will use the term *tile* and *active tile* interchangeably.

Assemblies. An assembly is a set of active tiles whose centers are located at integer coordinates, and no two tiles in the set are at the same location. For an assembly A, define the weighted graph $G_A = (V, E)$ such that $V = A$, and for any pair of tiles $a, b \in V$, the weight of edge (a, b) is defined to be 0 if a and b

do not have an overlapping face, and if a and b have overlapping faces d_a and d_b, the weight is defined to be $|\{i : state(a_{d_a,i}) = state(b_{d_b,i}) = on\}|$. That is, the weight of two adjacent tiles is the total number of matching glue types from a and b's overlapping edges that are both in state on. Conceptually, each such pair of equal, on glues represents a bond between a and b and thus increases the bonding strength between the tiles by 1 unit. For a positive integer τ, an assembly A is said to be τ-*stable* if the min-cut of the bond graph G_A is at least τ. For an assembly A, there is an associated pattern $p(A)$ defined by mapping the labels of each tile to corresponding lattice positions, and mapping the null label to lattice positions corresponding to locations not covered by the assembly.

Reactions. A reaction is an ordered pair of sets of assemblies. Conceptually, a reaction (A, B) represents the assemblies of set A replacing themselves with the assemblies in set B. For a reaction $r = (A, B)$, let r_{in} denote the set A, and r_{out} denote the set B. For a set of reactions R, let $R_{in} = \bigcup_{r \in R} r_{in}$ and $R_{out} = \bigcup_{r \in R} r_{out}$.

A reaction (A, B) is said to be *valid* for a given temperature τ if it is either a *break*, *combination*, or *glue-flip* reaction as defined below:

- **Break reaction.** A reaction $(A = \{a\}, B = \{b_1, b_2\})$ with $|A| = 1$ and $|B| = 2$ is said to be a break reaction if the bond graph of a has a cut of strength less than τ that separates a into assemblies b_1 and b_2.
- **Combination reaction.** A reaction $(A = \{a_1, a_2\}, B = \{b\})$ with $|A| = 2$ and $|B| = 1$ is said to be a combination reaction if a_1 and a_2 are *combinable* into assembly b (see definition below).
- **Glue-flip reaction.** A reaction $(A = \{a\}, B = \{b\})$ with $|A| = 1$ and $|B| = 1$ is said to be a glue-flip reaction if assembly b can be obtained from assembly a by changing the state of a single glue slot x in b to either on from latent if x has queued command on, or off from on or latent if x has queued command off. Note that transitions among latent, on, and off form acyclic graph with sink state off, implying glues states can be adjusted at most twice. This models the "fire once" property of signals.

Two assemblies a_1 and a_2 are said to be *combinable* if a_1 and a_2 can be translated such that a_1 and a_2 have no overlapping tile bodies, but have at least τ on, matching glues connecting tiles from a_1 to tiles from a_2. Given this translated pair of assemblies, consider the product assembly b to be the assemblies a_1 and a_2 merged with the queued commands for each glue slot set according to the specifications of the glue functions for each tile with newly bonded on glues along the cut between a_1 and a_2. In this case we say a_1 and a_2 are *combinable* into assembly b. See Figure 1 for example reactions and [9] for a more detailed presentation of the model.

Batches. A batch is a multi-set of assemblies in which elements may have either a non-negative integer multiplicity or an ∞ multiplicity. A batch B is said to be τ-*transitional* to a batch B' if the application of one of the break, combination, or transition rules at temperature τ can be applied to B to get B'. A *batch sequence* for some temperature τ is any sequence of batches $\langle a_1, \ldots a_r \rangle$ such that a_i is τ-transitional to a_{i+1} for each i from 1 to $r - 1$.

Signal Tile System. A signal tile system is an ordered pair (B, τ) where B is a batch referred to as the *initial seed* batch, and τ is a positive integer referred to as the temperature of the system. Any batch B' is said to be *producible* by (B, τ) if there exists a valid assembly sequence $\langle B_1, \ldots, B_r \rangle$ with respect to temperature τ such that $B' = B_r$ and $B = B_1$, i.e., B' is reachable from B by a sequence of τ-transitions.

2.3   Exponential Replication

Our first primary definition towards the concept of exponential replication defines a transition between batches in which multiple reactions may occur in parallel to complete the transition. By counting the number of such parallelized transitions we are able to define the number of time steps taken for one batch to transform into another, and in turn can define the concept of exponential replication.

However, to avoid reliance on highly unlikely reactions, we parameterize our definition with a positive integer c which dictates that any feasible combination reaction should involve at least one combinate with at least multiplicity c. By doing so, our exponential replication definition will be able to exclude systems that might rely on the highly unlikely combination of low concentration combinates (but will still consider such reactions in a worst-case scenario by requiring the subsequent monotonicity requirement). The following definition formalizes this concept.

Definition 1 ((τ, c)-transitional distance). *We say a batch B is (τ, c)-transitional to a batch B', with notation $B \rightarrow_{\tau, c} B'$, if there exists a set of reactions $R = COMBO \bigcup BREAK \bigcup FLIP$, where COMBO, BREAK, and FLIP partition R into the combination, break, and flip type reactions, such that:*

1. *$B - R_{in}$ is defined and $B' = B - R_{in} + R_{out}$.*
2. *For each $(\{x, y\}, \{z\}) \in COMBO$, the multiplicity of either x or y in $B - R_{in}$ is at least c.*

Further, we use notation $B \rightarrow^t_{\tau, c} B'$ if there exists a sequence $\langle B_1, \ldots, B_t \rangle$ such that $B_1 = B$, $B_t = B'$, and $B_i \rightarrow_{\tau, c} B_{i+1}$ for i from 1 to $t - 1$. We define the (τ, c)-transitional distance from B to B' to be the smallest positive integer t such that $B \rightarrow^t_{\tau, c} B'$.

Our next primary concept used to define exponential replication is the concept of monotonicity which requires that a sequence of batches (regardless of how likely) has the property that each subsequent batch in the sequence is at least as close (in terms of (τ, c)-transition distance) to becoming an element of a given goal set of batches as any previous batch in the sequence.

Definition 2 (Monotonicity). *Let B be a batch of assemblies, τ a positive integer, and G a set of (goal) batches. We say B grows monotonically towards G at temperature τ if for all temperature τ batch sequences $\langle B, \ldots, B' \rangle$, if $B \rightarrow_{\tau,c}^{t} g$ for some $g \in G$, then $B' \rightarrow_{\tau,c}^{t'} g'$ for some $g' \in G$ and $t' \leq t$.*

Note that g' in the above definition may differ from g. This means that B is not required to grow steadily towards any particular element of G, but simply must make steady progress towards becoming an element of G.

We now apply the concepts of (τ, c)-transition distance and monotonicity to define exponential replication of patterns. Informally, an STAM system is said to replicate the pattern of an assembly a if it is always guaranteed to have a logarithmic (in n) sequence of *feasible* transitions that will create at least n copies of a shape with a's pattern for any integer n. Further, to ensure that the system makes steady progress towards the goal of n copies, we further require the property of *monotonicity* which states that the number of transitions needed to attain the goal of n copies never increases, regardless of the sequence of reactions.

Definition 3 (Exponential Replication). *Let B_p^n denote the set of all batches which contain an n or higher multiplicity assembly with pattern p. A system $T = (B, \tau)$ exponentially replicates the pattern of assembly a if for all positive integers n and c:*

1. *$B \bigcup a \rightarrow_{\tau,c}^{t} B'$ for some $B' \in B_{p(a)}^{n}$ and $t = O(\mathit{poly}(|a|)\log(cn))$.*
2. *B grows monotonically towards $B_{p(a)}^{n}$.*

Given the concept of a system replicating a specific assembly, we now denote a system as a general *exponential replicator* if it replicates all patterns given some reasonable format that maps patterns to input assemblies. Let M denote a mapping from rectangular patterns over some alphabet ϕ to assemblies with the property that for any rectangular pattern w over ϕ, it must be that 1) $w = p(M(w))$ (The assembly representing pattern w must actually have pattern w), 2) all tiles in $M(w)$ with the same non-null label are the same active tile up to translation, and 3) the number of tiles in $M(w)$ is at most an additive constant larger than the size of w. Such a mapping is said to be a *valid format mapping* over ϕ. We now define what constitutes an exponential pattern replicator system.

Definition 4 (Exponential Replicator). *A system $T = (B, \tau)$ is an exponential pattern replicator for patterns over ϕ if there exists a valid format mapping M over ϕ such that for any rectangular pattern w over ϕ, $T = (B, \tau)$ exponentially replicates $M(w)$.*

3   Replication of 2D Patterns in Two Dimensions

We first informally discuss the mechanism for replication of 2D patterns in two dimensions with the tileset shown in Figure 3. Note that the same mechanism can be used for the replication of 2D patterns in three dimensions, and that in such a case, the disassembly and reassembly of the template may be omitted. For brevity, we do not show 2D pattern replication in three dimensions, as it is a trivial simplification of the process described in this paper. The replication process described here can be summarized in three phases. In the first phase, *template disassembly*, a template R containing some pattern over some alphabet ϕ is combined with the tile set that can replicate R. Initially, an inverted staircase cooperatively grows along the west face of R (Fig. 2, Phase 1). The effect of this tile growth is that each row of the original assembly R has a unique number of tiles appended to its west side. These appendages are used in reassembly later in the replication process. As the inverted staircase structure grows, rows of the original template are signaled to detach from each other. In Phase 2, the detached rows of the input assembly are available to serve as templates for the formation of *non-terminal replicates* (Fig. 2, Phase 2). Two types of replicate products are formed: *terminal replicates* (*tr*) and *non-terminal replicates* (*ntr*). While the pattern of each type of replicate is identical to that of the parent, each replicate type serves a different function. *Non-terminal replicates* may catalyze the formation of more product while *terminal* replicates serve as a final product and may not catalyze the formation of more product. After formation, this first generation of non-terminal replicates detach from the parent and enter Phase 3. In Phase 3, each *non-terminal replicate* may serve as a template for the formation of another *ntr* and a *tr* concurrently. The *tr* detaches from the parent upon completion and assembles, along with other terminal replicates, into a copy of R. Also during Phase 3, when the new *non-terminal replicate* is fully formed, it may detach from the parent and begin producing replicates (Fig. 2, Phase 3).

Theorem 1. *For any alphabet ϕ, there exists an exponential pattern replicator system $\Gamma = (T, 2)$ for patterns over ϕ. Furthermore, the seed batch T consists of $O(\phi)$ distinct singleton active tile types with a total of $O(\phi)$ unique glues.*

We prove this theorem by construction and present such a tile set below. The 12 active tile types which comprise T are depicted in Figure 3d-f. Note that the input pattern itself is not included in T. The input pattern to be replicated is of the form shown in Figure 3c, and this, together with T, comprises the initial seed batch. The pattern is mapped onto this input via the composition of the *Label* signal tiles. Figure 3a shows the tile types for a binary alphabet, while Figure 3b shows the tile type for some a_i of alphabet ϕ which consists of elements $a_1, a_2, \ldots a_\phi$.

Template disassembly and First Generation of Replicates. Upon addition of the template assembly R to the replicating tile set T, an inverted staircase forms on the west side of R (Fig. 4a). Concurrently, an end cap attaches to the east side of R. Note that while the east-side end caps are attaching to R, it is possible that an *ntr* tile type (white) found in Fig. 3e may attach to the north side of an end cap, blocking the attachment of an endcap to a row. This does not adversely affect replication, because given a temperature of 2, the template will still disassemble and the end cap may attach to rows lacking end caps following this event. Also, given that the north face label glues a_i' of the northernmost template row are

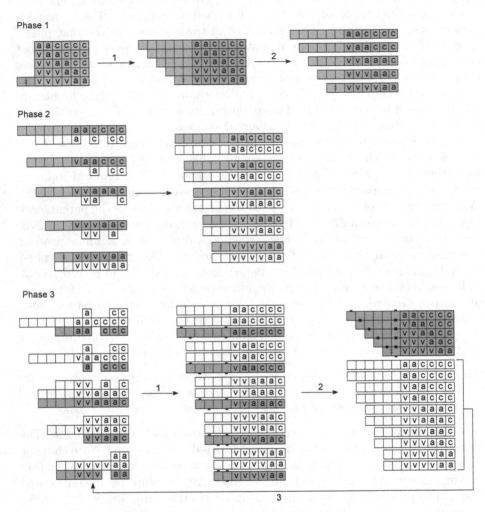

Fig. 2. The three phases shown above provide a general overview of the replication system described in this paper. In Phase 1, an inverted staircase (green) cooperatively grows along the west face of the pattern to be replicated (blue). Upon completion of the staircase, the assembly splits into distinct rows. In Phase 2, each of these distinct rows serves as a template for the production of a *non-terminal replicates* (*ntr*), shown in white, which has an identical pattern. In Phase 3, these *ntr*s serve as templates for the production of identical *ntr*s and *terminal replicates* (*tr*), which are shown in orange. The *tr*s reassemble to form a copy of the original pattern while the *ntr*s continue to serve as templates for the production of more *tr*s and *ntr*s.

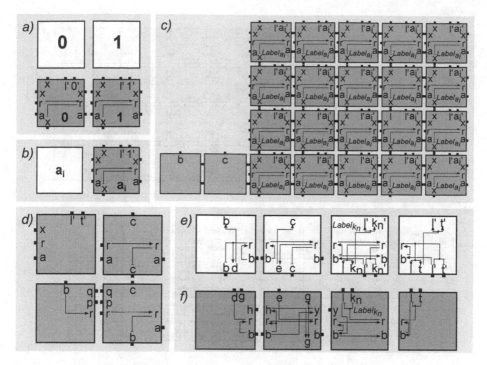

Fig. 3. a) Input assemby tile types for a binary alphabet b) The tile type for some a_i of alphabet ϕ which consists of elements $a_1, a_2, \ldots a_\phi$ c) General form of template to be replicated R d) Tiles involved in inverted staircase construction and disassembly of the original template. e) Tiles involved in formation of *non-terminal replicates*. f) Tiles involved in formation of *terminal replicates*.

exposed, it is possible for this row to begin replicating immediately. In fact, this is necessary for the row immediately below the northernmost row to detach. Any row s of R may release the row below it by turning off its south face glues (Fig. 4b). This can occur only if the row above s has activated the b glue on the westernmost tile of s. A signal is then propagated from west to east in row s via glue r and all south-face glues of s are turned off.

Following R disassembly, label glues a_i' are exposed on the north face of each row of the input assembly. Tiles involved in ntr formation (white) may attach along the north face of the template row (blue/green) (Fig. 5a). Following attachment, west face b glues are turned on. Once the westernmost $Label$ tile has attached, appendage tiles may cooperatively attach, sending a signal via b glues from west to east and turning on r glues. (Fig. 5b). After the westernmost appendage tile has attached, a signal is propagated from west to east via glue r queueing label glues a_i' on the south face of the new ntr to turn off, thus detaching the ntr from its parent (Fig. 5c). Label glues a_i are also queued on. These glues serve to generate a terminal replicate (tr) on the south face of the ntr (Fig. 5d). Following the detachment of the ntr and the parent template, the

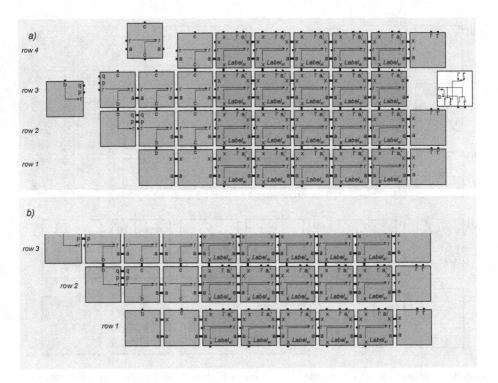

Fig. 4. a) Growth of inverted staircase along the west face of R and a cap on the east face of R. b) Row 1 is released after the b glue is activated on the westernmost tile of Row 2.

parent template is available to generate another *ntr*, while the first-generation *ntr* is immediately available to generate a *tr*.

Exponential Replication and Reassembly. After the formation of the first-generation *ntr*, replication is free to proceed exponentially. Glues on the south face of the *ntr* may bind label tiles from the *tr* tile set (Fig. 6a). Upon binding, *b* glues are turned *on* on the west face of the *tr* label tiles, allowing for the binding of appendage tiles on the western side of the growing *tr* assembly. Upon binding of the first appendage tile (Fig. 6b), a signal is propagated through the *tr* via glue *b* and *r* glues on the west faces of the *tr* tiles. After the next appendage tile binds, the *y* glue on the tile adjacent to it is activated, which activates two *g* glues on the north and south faces of the easternmost appendage tile (Fig. 6c). These *g* glues will assist in proper reassembly of each row into a correct copy of the template R. Also note that upon binding a *tr* tile, label glues a'_i on the north face of the *ntr* are turned *on*. This allows for synthesis of a new *ntr* on the north side of the parent *ntr* while a new *tr* is being formed on the south face. The synthesis of a new *ntr* from a parent *ntr* is not described in detail here, as it is very similar to the process described in Figure 5. Upon attachment of

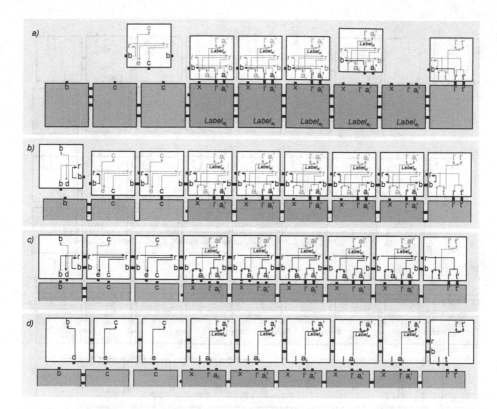

Fig. 5. The above sequence outlines details of the production of *non-terminal replicates*. For clarity, glues turned *off* and signals previously executed are not shown.

the westernmost appendage tile, a north face g glue of the tr is turned *on* as well as a south face g glue on the tile immediately adjacent to it. Additionally, a signal is propagated from west to east along the tr via glue r and the north face glues of the tr are turned *off*. The tr then detaches from the parent ntr (Fig. 6d) and is available for reassembly into a copy of the original template R while the parent ntr is available to produce a new ntr on its north face and a new tr on its south face. The alignment of g glues enables the proper reassembly of the *terminal replicates* into a copy of R (Fig. 7).

The detachment of the inverted staircase is not described here. If a signal cascade were designed such that upon the complete assembly of a copy of the original template pattern, the inverted staircase detached, it would be considered a waste product. The number of these waste assemblies would grow proportionally to the number of replicates of R. Similarly, if the replication process were somehow halted, and the copies of R harvested, the $ntrs$ might also be considered waste. These, too, would have grown proportionally to the copies of R.

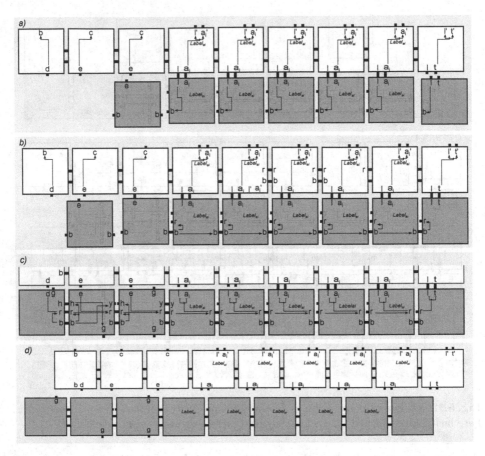

Fig. 6. The above sequence shows details of the formation of *terminal replicates*. For clarity, glues turned *off* and signals executed during template disassembly are not shown.

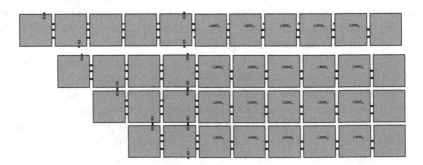

Fig. 7. *Terminal replicates* reassemble into a copy of R

4   Future Work

The results of this paper provide several directions for future work. One interesting problem is the replication of shapes in the STAM, or more specifically, patterned shapes. One might imagine a mechanism similar to the one presented in this paper but where the growth of the inverted staircase is preceded by a "rectangularization" of the shape to be replicated. The replication of a cuboid is conceivable by extending the mechanism of template disassembly and reassembly presented in Section 3 to three dimensions where layers of the cuboid might be separated, replicated, and the replicates reassembled. Precise replication of a certain number of copies could also be possible, as was considered in [1].

Another direction for future work is studying the extent to which staged self-assembly systems can be simulated by non-staged active self-assembly systems such as the signal tile model. In [3] efficient staged algorithms are developed to assemble linear structures, while a signal tile system achieves a similar result in [9]. Shape replication through stages and RNA based tiles are used to replicate general shapes in [1], while this paper and future work suggests similar results may be obtained with signal tiles. Can the complexity of the mixing algorithm of a staged assembly algorithm be encoded into a signal tile system of similar complexity? As a first step towards such a simulation we might consider the case of 1D assemblies. Can the efficient construction of labeled linear assemblies through staging shown in [4] be efficiently simulated with a signal tile system?

A final direction for future work involves the simulation of the signal tile model through a passive model of self-assembly such as the abstract or two-handed tile assembly model [2]. Recent work has shown how restricted classes of signal tile systems can be simulated by passive 3D systems [5]. The ability for signal tile systems to perform *fuel-efficient* computation was shown to be achievable within passive 2D tile assembly given the added power of negative force glues [14]. Is it possible to simulate *any* signal tile system with the use of negative glues? Can this be done in 2D?

Acknowledgements. We would like to thank Jennifer Padilla for helpful discussions regarding the current state of experimental DNA implementations of signal tile systems.

References

1. Abel, Z., Benbernou, N., Damian, M., Demaine, E., Demaine, M., Flatland, R., Kominers, S., Schweller, R.: Shape replication through self-assembly and RNase enzymes. In: SODA 2010: Proceedings of the Twenty-first Annual ACM-SIAM Symposium on Discrete Algorithms, Austin, Texas. Society for Industrial and Applied Mathematics (2010)
2. Cannon, S., Demaine, E.D., Demaine, M.L., Eisenstat, S., Patitz, M.J., Schweller, R., Summers, S.M., Winslow, A.: Two hands are better than one (up to constant factors). Arxiv preprint arXiv:1201.1650 (2012)

3. Demaine, E.D., Demaine, M.L., Fekete, S.P., Ishaque, M., Rafalin, E., Schweller, R.T., Souvaine, D.L.: Staged self-assembly: nanomanufacture of arbitrary shapes with $O(1)$ glues. Natural Computing 7(3), 347–370 (2008)
4. Demaine, E.D., Eisenstat, S., Ishaque, M., Winslow, A.: One-dimensional staged self-assembly. In: Cardelli, L., Shih, W. (eds.) DNA 17. LNCS, vol. 6937, pp. 100–114. Springer, Heidelberg (2011)
5. Hendricks, J., Padilla, J.E., Patitz, M.J., Rogers, T.A.: Signal transmission across tile assemblies: 3D static tiles simulate active self-assembly by 2D signal-passing tiles. In: Soloveichik, D., Yurke, B. (eds.) DNA 2013. LNCS, vol. 8141, pp. 90–104. Springer, Heidelberg (2013)
6. Kao, M.-Y., Schweller, R.T.: Reducing tile complexity for self-assembly through temperature programming. In: SODA 2006: Proceedings of the 17th Annual ACM-SIAM Symposium on Discrete Algorithms, pp. 571–580 (2006)
7. Marchal, P.: John von neumann: The founding father of artificial life. Artificial Life 4(3), 229–235 (1998)
8. Padilla, J.E., Liu, W., Seeman, N.C.: Hierarchical self assembly of patterns from the Robinson tilings: DNA tile design in an enhanced tile assembly model, Natural Computing (online first August 17, 2011)
9. Padilla, J.E., Patitz, M.J., Pena, R., Schweller, R.T., Seeman, N.C., Sheline, R., Summers, S.M., Zhong, X.: Asynchronous signal passing for tile self-assembly: Fuel efficient computation and efficient assembly of shapes. In: Mauri, G., Dennunzio, A., Manzoni, L., Porreca, A.E. (eds.) UCNC 2013. LNCS, vol. 7956, pp. 174–185. Springer, Heidelberg (2013)
10. Patzke, V., von Kiedrowski, G.: Self-replicating sytems. ARKIVOC 5, 293–310 (2007)
11. Paul, N., Joyce, G.F.: A self-replicating ligase ribozyme. PNAS 99(120), 12733–12740 (2002)
12. Reif, J.H., Sahu, S., Yin, P.: Complexity of graph self-assembly in accretive systems and self-destructible systems. In: Carbone, A., Pierce, N.A. (eds.) DNA11. LNCS, vol. 3892, pp. 257–274. Springer, Heidelberg (2006)
13. Schulman, R., Yurke, B., Winfree, E.: Robust self-replication of combinatorial information via crystal growth and scission. PNAS 109(17), 6405–6410 (2012)
14. Schweller, R., Sherman, M.: Fuel efficient computation in passive self-assembly. In: Proceedings of the 24th Annual ACM-SIAM Symposium on Discrete Algorithms (SODA 2013), New Orleans, Louisiana, pp. 1513 – 1525 (2013)
15. Szathmary, E., Gladkih, I.: A self-replicating hexadeoxynucleotide. Journal of Theoretical Biology 138(1), 55–58 (1989)
16. Tjivikua, T., Ballester, P., Rebek Jr., J.: Self-repllicating system. J. Am. Chem. Soc. 112(3), 1249–1250 (1990)
17. von Kiedrowski, G.: A self-replicating hexadeoxynucleotide. Angewandte Chemie International Edition in English 25(10), 932–935 (1986)
18. Winfree, E.: Algorithmic self-assembly of DNA, Ph.D. thesis, California Institute of Technology (June 1998)
19. Zielinski, W., Orgel, L.: Autocatalytic synthesis of a tetranucleotide analogue. Nature 327, 346–347 (1987)

Modular Verification of DNA Strand Displacement Networks via Serializability Analysis

Matthew R. Lakin[1], Andrew Phillips[2], and Darko Stefanovic[1,3]

[1] Department of Computer Science, University of New Mexico, NM 87131, USA
[2] Biological Computation Group, Microsoft Research, Cambridge, CB1 2FB, UK
[3] Center for Biomedical Engineering, University of New Mexico, NM 87131, USA
{mlakin,darko}@cs.unm.edu, andrew.phillips@microsoft.com

Abstract. DNA strand displacement gates can be used to emulate arbitrary chemical reactions, and a number of different schemes have been proposed to achieve this. Here we develop modular correctness proofs for strand displacement encodings of chemical reaction networks and show how they may be applied to two-domain strand displacement systems. Our notion of correctness is serializability of interleaved reaction encodings, and we infer this global property from the properties of the gates that encode the individual chemical reactions. This allows correctness to be inferred for arbitrary systems constructed using these components, and we illustrate this by applying our results to a two-domain implementation of a well-known approximate majority voting system.

1 Introduction

The behaviour and kinetics of arbitrary chemical reaction networks can be emulated by collections of DNA strand displacement gates [1,2]. A number of schemes for encoding reactions using strand displacement gates have been proposed, such as four-domain [2], three-domain [3,4] and two-domain schemes [5]. A common feature of these gates is that they emulate a single-step reaction using a sequence of multiple reactions. These additional steps introduce more opportunities for errors in the design, which may include subtle concurrency bugs that only manifest themselves when a gate is used in a particular context. Therefore it is desirable to develop formal proofs that a given strand displacement gate design is a correct implementation of the desired chemical reactions.

Two-domain encoding schemes are attractive candidates for experimental implementation because they use simple strands and gates without overhangs. However, this necessitates additional intermediate steps in the emulation of the reaction, such as garbage collection steps that convert leftover species into unreactive waste to prevent them slowing down certain reactions. Since a larger number of steps are often needed to encode a given reaction, it is typically less obvious that the design is correct. In previous work we have explored the use of probabilistic model checking for the verification of two-domain strand displacement systems [6]. However, in that work we could only verify particular populations of species, and this was severely limited by the explosion in the size of the state space as the sizes of the species populations increased.

Here we introduce a framework for verification of DNA strand displacement emulations of chemical reaction networks. We adopt a modular approach to proving correctness, by showing that if all components of the system satisfy certain properties then the

D. Soloveichik and B. Yurke (Eds.): DNA 2013, LNCS 8141, pp. 133–146, 2013.
© Springer International Publishing Switzerland 2013

whole system may be deduced to be correct in a well-defined sense. Our approach is
inspired by the concept of *serializability* from database theory [7], which requires that
interleaved concurrent updates to a database must be equivalent to some serial schedule of those updates. We consider a composition of reaction encodings to be correct if
all interleavings of their reactions are causally equivalent to some serial schedule, in
which there is no interleaving between the various reaction encodings. We use simple
rewriting rules on reaction traces to serialize them, and apply our technique to the verification of two-domain DNA strand displacement reaction gates [5,6]. We propose serializability as a reasonable notion of correctness for DNA strand displacement reaction
gates because serialized executions of encodings can be directly related to executions of
the underlying reactions. Gate designs that are not serializable may display erroneous
behaviours that do not correspond to possible behaviours of the underlying reactions,
because of unwanted crosstalk between gates. Our correctness criteria will allow us to
prove that gate designs do not have such problems.

2 Preliminaries

We now make some preliminary mathematical definitions that will be used throughout
the paper. Let $\mathbb{N}$ denote the set of natural numbers, including zero. Following the notation of [8], given a set X, we write $\mathbb{N}^X$ for the set of multisets over X, defined as the
set of all functions $f : X \to \mathbb{N}$. By convention we use upper-case boldface symbols
for multisets and upper-case italics for sets. We may write multisets explicitly using
the notation $\{x_1 = n_1, \ldots, x_k = n_k\}$, where n_i is the count associated with the corre-
sponding x_i. For multisets $\mathbf{A}, \mathbf{B} \in \mathbb{N}^X$ we write $\mathbf{A} \circledast \mathbf{B}$ to mean that $(\mathbf{A}(x)) \circledast (\mathbf{B}(x))$
for all $x \in X$, where $\circledast$ is any binary relational operator, for example $\leq$. Similarly, we
define arithmetic operations on multisets so that $(\mathbf{A} \pm \mathbf{B})(x) = (\mathbf{A}(x)) \pm (\mathbf{B}(x))$ for
all $x \in X$. For subtraction we require that $\mathbf{B} \leq \mathbf{A}$ to avoid negative multiplicities. If
$x \in X$ and $n \in \mathbb{N}$, we write $n \cdot x$ for the multiset $\mathbf{A} \in \mathbb{N}^X$ such that $\mathbf{A}(x) = n$ and
such that $\mathbf{A}(x') = 0$ for $x' \in X$ where $x' \neq x$. We now define some key concepts.

Definition 1 (Chemical reaction networks). *A chemical reaction network (CRN) is a
pair* (X, R), *where* X *is a set of chemical species and* R *is a set of chemical reactions
over* X. *A chemical reaction has the form* $\mathbf{R} \to \mathbf{P}$, *where* $\mathbf{R}, \mathbf{P} \in \mathbb{N}^X$ *represent the
reactants and products of the reaction, respectively. If* $r = (\mathbf{R} \to \mathbf{P})$ *then we let* $r^{-1} =
(\mathbf{P} \to \mathbf{R})$ *and observe that* $(r^{-1})^{-1} = r$. *Note that we do not consider reaction rates
at all in this paper. For well-formedness of reactions we will stipulate that* $\mathbf{R} \neq \mathbf{P}$, *and
for well-formedness of CRNs we require that all constituent reactions are well-formed.
Henceforth we assume that all CRNs are well-formed.*

Definition 2 (CRN states and reduction). *A state* $\mathbf{S}$ *of a CRN* $\mathcal{C} = (X, R)$ *is just a
multiset drawn from* $\mathbb{N}^X$. *A reaction* $r = (\mathbf{R} \to \mathbf{P}) \in R$ *is enabled in state* $\mathbf{S}$ *if* $\mathbf{R} \leq \mathbf{S}$,
written $\mathbf{S} \vdash_{\mathcal{C}} r$. *Furthermore, if* $\mathbf{S}' = \mathbf{S} - \mathbf{R} + \mathbf{P}$ *then we write* $\mathbf{S} \xrightarrow{r} \mathbf{S}'$ *to indicate
that applying the reaction* r *to* $\mathbf{S}$ *results in* $\mathbf{S}'$.

Definition 3 (CRN traces). *Given a CRN* $\mathcal{C} = (X, R)$, *a trace* τ *is an ordered list*
$[r_1, \ldots, r_i, \ldots]$ *of elements of* R. *Traces may be finite or infinite, and we write length*(τ)

for the length of the finite trace τ. *We write* $\text{Traces}(\mathcal{C})$ *for the set of all traces that may be generated using reactions from R. We write* $\tau_1 : \tau_2$ *for the trace obtained by concatenating* τ_1 *and* τ_2, *and* ϵ *to denote the empty trace (i.e.,* $\text{length}(\epsilon) = 0$).

Definition 4 (Valid reductions). *A pair* $(\mathbf{S}, \tau)$ *is a* valid reduction *of a CRN* $\mathcal{C}$, *written* $\mathbf{S} \vdash_{\mathcal{C}} \tau$, *if* $\tau \in \text{Traces}(\mathcal{C})$ *and either (i)* $\tau = \epsilon$ *or (ii)* $\tau = [r] : \tau'$ *and* $\mathbf{S} \xrightarrow{r} \mathbf{S}'$ *such that* $\mathbf{S}' \vdash_{\mathcal{C}} \tau'$ *also holds. If* τ *is finite, we write* $\text{final}_{\mathcal{C}}(\mathbf{S}, \tau)$ *for the final state of the trace, and say that* $\mathbf{S} \xrightarrow{\tau} \mathbf{S}'$ *if* $\mathbf{S}' = \text{final}_{\mathcal{C}}(\mathbf{S}, \tau)$.

Definition 5 (Reachable states). *A state* $\mathbf{S}'$ *is* reachable *from a state* $\mathbf{S}$ *under a CRN* $\mathcal{C} = (X, R)$ *if* $\mathbf{S} \xrightarrow{\tau} \mathbf{S}'$ *holds for some* $\tau \in \text{Traces}(\mathcal{C})$. *Furthermore, we say that a state* $\mathbf{S}'$ *is* universally reachable *from* $\mathbf{S}$ *under* $\mathcal{C}$ *if* $\mathbf{S}'$ *is reachable from every state that is reachable from* $\mathbf{S}$.

Definition 6 (Terminal states and traces). *A state* $\mathbf{S}$ *is* terminal *under a CRN* $\mathcal{C} = (X, R)$ *if no reaction* $r \in R$ *is enabled in* $\mathbf{S}$. *A* terminal trace *from a state* $\mathbf{S}$ *is any finite trace* $\tau \in \text{Traces}(\mathcal{C})$ *such that* $\text{final}_{\mathcal{C}}(\mathbf{S}, \tau)$ *is a terminal state under* $\mathcal{C}$.

Definition 7 (Reversible and irreversible reactions). *Given a CRN* $\mathcal{C} = (X, R)$, *a reaction* $r \in R$ *is* reversible *if the inverse reaction* r^{-1} *also appears in R, and it is* irreversible *if* $\mathbf{S} \xrightarrow{r} \mathbf{S}'$ *implies that* $\mathbf{S}$ *is not reachable from* $\mathbf{S}'$ *under* $\mathcal{C}$. *Note that it is* not *the case that every reaction is necessarily either reversible or irreversible in the above sense: for example, consider the CRN with reactions* $a \rightarrow b$, $b \rightarrow c$ *and* $c \rightarrow a$. *None of these reactions are reversible, but none of the reactions are irreversible either, because there is always a route back to the previous state via the other two reactions.*

3   Two-Domain DNA Strand Displacement Gates

We cannot develop a modular verification framework without a common language in which to formalize the system components: this allows us to check for unwanted crosstalk between modules in a uniform way. We will use the DSD language for formalizing DNA strand displacement systems [4,9]. Here we present a subset of the syntax and reactions needed for the two-domain gates in this paper: see [4] for full definitions.

The syntax and graphical notation for two-domain DSD systems is presented in Table 1. We write `t^`, `u^`, etc., for *toehold domains*, which are short enough to hybridize reversibly (shown in black), and `a`, `b`, `x`, `y`, etc., for *recognition domains*, which are long enough to hybridize irreversibly (shown in grey). We use the asterisk to denote the Watson-Crick complement of a particular domain, and assume that the domains are non-interfering, i.e., `x` will only hybridize with `x*`. Single strands S may be *signals* `<t^ x>` or *cosignals* `<x u^>`. Following [6] and [5], we extend the basic two-domain syntax with extended strands to enable irreversible product release, by including extended strands of the form `<t^ x y>` and `<x y u^>`. Finally, certain reactions can produce waste strands of the form `<x>`, which are unreactive as they have no toehold to interact with other species.

Figure 1 presents the set of possible reactions between two-domain DNA strands and gates presented in Table 1, according to the *Infinite* DSD semantics [4]. In reaction *(i)*, a

Table 1. DSD syntax and graphical representation of two-domain DNA species. Here, G1 and G2 stand for arbitrary, non-empty gate structures.

Species category	DSD syntax	Graphical notation
Strands, S		
Signal strand	`<t^ x>`	
Cosignal strand	`<x u^>`	
Extended signal strand	`<t^ x y>`	
Extended cosignal strand	`<x y u^>`	
Inert waste strand	`<x>`	
Gates, G		
Exposed toehold gate segment	`{t^*}`	
Double-stranded signal gate segment	`[t^ x]`	
Double-stranded cosignal gate segment	`[x u^]`	
Collector gate segment	`[x]`	
Segment concatenation	`G1:G2`	Lower strands joined

signal strand is consumed by a gate via an exposed complementary toehold, producing a new gate with a different exposed toehold and a free cosignal strand. Note that this reaction is reversible, since the cosignal can react with the product gate to release the original signal into solution. Reactions *(ii)* and *(iii)* show irreversible consumption of a cosignal and a signal respectively, by gates containing the appropriate combination of a collector segment and an exposed complementary toehold, sealing off the toehold and releasing an inert waste strand into solution. Finally, reactions *(iv)* and *(v)* show irreversible consumption of an extended signal and an extended cosignal respectively. These reactions seal off a toehold and release an inert waste strand *and* an output strand, which is either a signal or a cosignal. (It is also possible to irreversibly consume an extended strand using two neighbouring collector segments without releasing a signal or cosignal, but we do not consider this because gate designs typically do not require it.) The reactions from Figure 1 are all either reversible or irreversible in the sense of Definition 7, and they are all bimolecular reactions that involve one gate and one strand as reactants. We ignore unproductive reactions [4] in which a toehold binds but cannot initiate a subsequent branch migration reaction.

4   Modular Chemical Reaction Encodings

The goal of this paper is to verify that a CRN involving DNA strand displacement reactions correctly encodes a particular CRN of interest. We specify CRN encodings in a modular way, by defining subsystems which each encode a particular chemical reaction, and composing these to form a single CRN. We refer to the species and reactions of the CRN being encoded as *formal* species and reactions, and let α range over formal

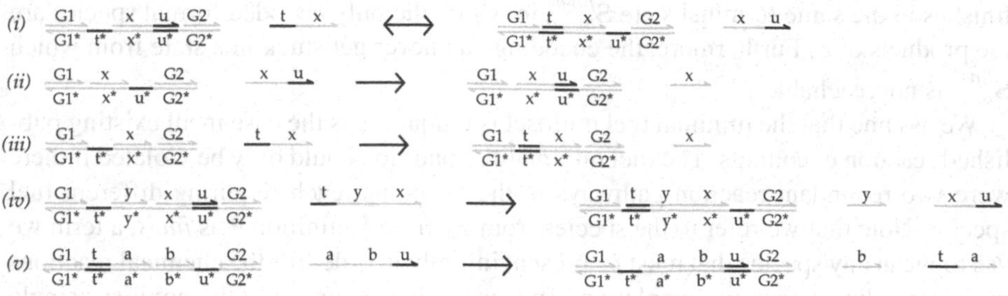

(i) $\frac{G1 \quad x \quad u \quad G2}{G1^* \quad t^* \quad x^* \quad u^* \quad G2^*}$ t x $\longleftrightarrow$ $\frac{G1 \quad t \quad x \quad G2}{G1^* \quad t^* \quad x^* \quad u^* \quad G2^*}$ x u

(ii) $\frac{G1 \quad x \quad G2}{G1^* \quad x^* \quad u^* \quad G2^*}$ x u $\longrightarrow$ $\frac{G1 \quad x \quad u \quad G2}{G1^* \quad x^* \quad u^* \quad G2^*}$ x

(iii) $\frac{G1 \quad x \quad G2}{G1^* \quad t^* \quad x^* \quad G2^*}$ t x $\longrightarrow$ $\frac{G1 \quad t \quad x \quad G2}{G1^* \quad t^* \quad x^* \quad G2^*}$ x

(iv) $\frac{G1 \quad y \quad x \quad u \quad G2}{G1^* \quad t^* \quad y^* \quad x^* \quad u^* \quad G2^*}$ t y x $\longrightarrow$ $\frac{G1 \quad t \quad y \quad x \quad G2}{G1^* \quad t^* \quad y^* \quad x^* \quad u^* \quad G2^*}$ y x u

(v) $\frac{G1 \quad t \quad a \quad b \quad G2}{G1^* \quad t^* \quad a^* \quad b^* \quad u^* \quad G2^*}$ a b u $\longrightarrow$ $\frac{G1 \quad a \quad b \quad u \quad G2}{G1^* \quad t^* \quad a^* \quad b^* \quad u^* \quad G2^*}$ b t a

Fig. 1. Two-domain DNA strand displacement reactions employed in this paper. Here, G1 and G2 stand for arbitrary, possibly-empty gate structures.

reactions and $\mathcal{F}$ over formal CRNs. We will encode formal species using the family of DSD species defined in Table 1, which interact via reactions derived from the rules in Figure 1.

Definition 8 (Species encodings). *For a set X of formal species, we define a bijective species map $\mathcal{M}$ which maps every $x \in X$ to a DSD species from Table 1. (As is standard, we assume that encoded formal species never interact directly.) In the case of two-domain systems, we fix a global toehold domain $t^\wedge$ and choose our species map such that $\mathcal{M}(x) = <t^\wedge \ x>$ for all $x \in X$, i.e., each formal species is encoded by a different domain. Writing x' and $\mathbf{S}'$ for encoded species and states respectively, we lift $\mathcal{M}$ (and the inverse mapping $\mathcal{M}^{-1}$) to operate on states by simply ignoring the species that are not present in their domain of definition, as follows.*

$$(\mathcal{M}(\mathbf{S}))(x') = \begin{cases} \mathbf{S}(\mathcal{M}^{-1}(x')) & \text{if } x' \in \text{image}(\mathcal{M}) \\ 0 & \text{otherwise.} \end{cases}$$

$$(\mathcal{M}^{-1}(\mathbf{S}'))(x) = \begin{cases} \mathbf{S}'(\mathcal{M}(x)) & \text{if } x \in \text{dom}(\mathcal{M}) \\ 0 & \text{otherwise.} \end{cases}$$

We will develop encodings of a formal CRN by constructing a *reaction encoding* $[\![\alpha]\!]$ for each constituent formal reaction α, as follows.

Definition 9 (Reaction encodings). *An encoding $[\![\alpha]\!]$ of a formal reaction $\alpha = (\mathbf{R}_\alpha \rightarrow \mathbf{P}_\alpha)$ is a multiset $\mathbf{F}_\alpha$ of fuel species from Table 1 such that, if we let $\mathbf{S}_\alpha^{init} = \mathcal{M}(\mathbf{R}_\alpha) + \mathbf{F}_\alpha$, then there exists a terminal state $\mathbf{S}_\alpha^{final}$ that is universally reachable from $\mathbf{S}_\alpha^{init}$ using the DSD reaction rules, and where $\mathcal{M}^{-1}(\mathbf{S}_\alpha^{init}) = \mathbf{R}_\alpha$ and $\mathcal{M}^{-1}(\mathbf{S}_\alpha^{final}) = \mathbf{P}_\alpha$. We also require that no $\mathbf{F} < \mathbf{F}_\alpha$ has the above properties, meaning that $\mathbf{F}_\alpha$ is the minimal amount of fuel needed to completely execute a single copy of the encoding. Finally, if $\mathbf{S}_\alpha^{init} \xrightarrow{\tau} \mathbf{S}_\alpha^{final}$ we say that the trace τ is an execution of $[\![\alpha]\!]$.*

Intuitively, a reaction encoding $[\![\alpha]\!]$ comprises the minimal amount of fuel $\mathbf{F}_\alpha$ which, when placed with the encoded reactants $\mathcal{M}(\mathbf{R}_\alpha)$, can execute the encoding of a *single instance* of the formal reaction α. The definition also requires that the encoding always finishes in the same terminal state S_α^{final}, in which the only encoded formal species are the products of α. Furthermore, the encoding can never get stuck in a state from which S_α^{final} is not reachable.

We assume that the minimal fuel multiset is unique, as is the case in all existing published reaction encodings. The fuel minimality condition could only be violated if there were two redundant reaction pathways in the encoding, each requiring different fuel species. Note that we refer to the species from F_α from Definition 9 as *fuels*, a term we use to mean any species that must be present initially in order for the chemical reactions in the encoding to run to completion. This encompasses not only the auxiliary single strands typically referred to as "fuels" in the literature, but also the gate complexes that must be present initially. Note that the requirement of a terminal state that is universally reachable implies that there can only be a single terminal state: if there were a second terminal state, then by definition the first would not be reachable from the second and hence would not be universally reachable.

Definition 10 (CRN encodings). *Suppose that $\mathcal{F}$ is a formal CRN that contains formal reactions $\alpha_1, \ldots, \alpha_n$ with corresponding encodings $[\![\alpha_1]\!], \ldots, [\![\alpha_n]\!]$. We use the individual reaction encodings to derive a CRN $\mathcal{E} = (X_\mathcal{E}, R_\mathcal{E})$ that encodes $\mathcal{F}$, by*

- *forming the set of initial species, which comprises all fuel species from the multisets $F_{\alpha_1}, \ldots, F_{\alpha_n}$ and all encoded formal species $\mathcal{M}(x)$, where x is mentioned in one of the formal reactions $\alpha_1, \ldots, \alpha_n$, and then*
- *recursively computing the set $X_\mathcal{E}$ of all reachable species and the set $R_\mathcal{E}$ of all possible reactions, using the reaction rules from Figure 1.*

We refer the reader to previous work which formally defined the reaction enumeration algorithm from the DSD compiler [4,10], which can be used to automate the process described in Definition 10.

The most basic correctness property of reaction encodings is that they are capable of emulating any valid trace of formal reactions. For a formal trace $\tau \in \mathrm{Traces}(\mathcal{F})$, we say that a trace $\tau' \in \mathrm{Traces}(\mathcal{E})$ in the encoded CRN is a *serial execution* of $\tau = [r_1, \ldots, r_n]$ if τ' can be decomposed into subtraces $\tau_1 : \cdots : \tau_n$ such that the ith subtrace τ_i is an execution of r_i. Since reaction encodings require fuel species to be present, any such statement must be predicated on the amount of fuel available in the system. Thus it is important to identify the minimal amount of fuel needed to emulate a given trace of formal reactions.

Lemma 1 (Fuel required for emulation). *Let $\tau = [\alpha_1, \ldots, \alpha_n] \in \mathrm{Traces}(\mathcal{F})$ be a finite trace of formal reactions, let S be a formal state such that $S \vdash_\mathcal{F} \tau$, and let $\tau_{ser} \in \mathrm{Traces}(\mathcal{E})$ be a serial execution of τ. Then, $(\mathcal{M}(S) + F) \vdash_\mathcal{E} \tau_{ser}$ iff $F \geq reqfuel(\tau)$, where the required fuel, $reqfuel(\tau)$, is defined as the sum of the fuel required by the encoding of each reaction in the formal trace, i.e., $reqfuel(\tau) \triangleq F_{\alpha_1} + \cdots + F_{\alpha_n}$.* $\square$

It follows from Lemma 1 that we can emulate any finite formal trace by creating an initial state with sufficient encoded species $\mathcal{M}(S)$ and fuels F so that the corresponding serial execution can be run, and there is an obvious connection between the serial execution and the formal trace.

Theorem 1 (Completeness). *Let $\tau \in \text{Traces}(\mathcal{F})$ be a finite formal trace and let $\mathbf{S}$ be a formal state. If $\mathbf{S} \vdash_{\mathcal{F}} \tau$ and $\mathbf{F} \geq \text{reqfuel}(\tau)$ then $(\mathcal{M}(\mathbf{S}) + \mathbf{F}) \vdash_{\mathcal{E}} \tau_{ser}$ for some $\tau_{ser} \in \text{Traces}(\mathcal{E})$ which is a serial execution of τ.* □

5 Soundness of CRN Encodings

The completeness result above only concerns one possible reduction trace of the encoding. In this section we prove a more involved *soundness* result, which shows that *all* possible reduction traces of the encoding are equivalent to a serial execution of some valid formal trace. This is a reasonable notion of correctness because this implies that every possible trace of the encodings can be causally related to some valid formal trace. To make this connection, we define a notion of *rewriting* on valid reaction traces, which allows reactions to be moved around and deleted from the trace if doing so preserves the causal relationships between reactions in the trace.

Definition 11 (Trace rewriting). *The trace rewriting relation is indexed by a CRN C and the starting state $\mathbf{S}$. We write $\mathbf{S} \vdash_C \tau \rightsquigarrow \tau'$ to mean that $\mathbf{S} \vdash_C \tau$ and that a derivation exists using the following inference rules.*

$$(\text{REFL}) \;\frac{}{\mathbf{S} \vdash_C \tau \rightsquigarrow \tau} \qquad\qquad (\text{TRANS}) \;\frac{\mathbf{S} \vdash_C \tau \rightsquigarrow \tau' \quad \mathbf{S} \vdash_C \tau' \rightsquigarrow \tau''}{\mathbf{S} \vdash_C \tau \rightsquigarrow \tau''}$$

$$(\text{CANCEL}) \;\frac{\mathbf{S} \xrightarrow{\tau_1} \mathbf{S}' \xrightarrow{\tau_2} \mathbf{S}'}{\mathbf{S} \vdash_C \tau_1{:}\tau_2{:}\tau_3 \rightsquigarrow \tau_1{:}\tau_3} \qquad (\text{SWAP}) \;\frac{\mathbf{S} \vdash_C \tau_1{:}\tau_3{:}\tau_2}{\mathbf{S} \vdash_C \tau_1{:}\tau_2{:}\tau_3{:}\tau_4 \rightsquigarrow \tau_1{:}\tau_3{:}\tau_2{:}\tau_4}$$

The (CANCEL) rule allows the subtrace τ_2 to be removed if its net effect is no change, and the (SWAP) rule allows two neighbouring subtraces τ_2 and τ_3 to be swapped if they may occur in either order. In the latter case, since executing a reaction trace is essentially a series of addition and subtraction operations on the species populations, which are commutative, it follows that $\tau_1{:}\tau_2{:}\tau_3$ and $\tau_1{:}\tau_3{:}\tau_2$ both produce the same final state. It is not hard to show that trace rewriting preserves validity and the final states of reductions, as stated below.[1]

Lemma 2. *If $\mathbf{S} \vdash_C \tau \rightsquigarrow \tau'$ then $\mathbf{S} \vdash_C \tau'$ and $\text{final}_C(\mathbf{S}, \tau) = \text{final}_C(\mathbf{S}, \tau')$.*

Note that it is *not* the case that any two valid traces from a given starting state must be trace-equivalent—indeed, this is the crux of our analysis. Proving soundness is challenging because we must show that *all* possible interleavings of the various reaction encodings can be rewritten to produce a serial execution of a valid formal trace. To obtain this result, we must place additional constraints on the reaction encodings which we will consider.

Definition 12 (Stratified chemical reaction networks). *If a state $\mathbf{S}'$ is reachable from $\mathbf{S}$ under C, we write $\Lambda(\mathbf{S}, \mathbf{S}')$ for the length of the shortest trace $\tau \in \text{Traces}(C)$ such*

---

[1] A proof sketch for Lemma 2 is included in the appendices, which can be downloaded from the first author's webpage.

that $S \xrightarrow{\tau} S'$. *Then, we say that the CRN* C *is* stratified *if, for any starting state* S_0 *and any states* S *and* S' *which are reachable from* S_0 *under* C, *it is the case that* $S \xrightarrow{r} S'$ *implies* $\Lambda(S_0, S') = \Lambda(S_0, S) \pm 1$.

We observe that reactions derived from Figure 1 always give rise to stratified CRNs, since the reversible reactions can only be reversed by executing the corresponding inverse reaction and the irreversible reactions all produce an inert species which prevents the system from returning to the previous state. Intuitively, this allows us to subdivide the transitions in our reaction encodings into *forward steps* that move away from the initial state S_0 (i.e., transitions $S \xrightarrow{r} S'$ where $\Lambda(S_0, S') = \Lambda(S_0, S) + 1$) and *backward steps* that move back towards the initial state S_0 (i.e., transitions $S \xrightarrow{r} S'$ where $\Lambda(S_0, S') = \Lambda(S_0, S) - 1$). We must also categorize the species involved in each reaction encoding according to their role: we require that the set of *all* species involved in the encoding $[\![\alpha]\!]$, denoted *species*$([\![\alpha]\!])$, can be partitioned into:

- *formals*$([\![\alpha]\!])$ (those species in the image of $\mathcal{M}$);
- *waste*$([\![\alpha]\!])$ (those species which are unreactive);
- *fuels*$([\![\alpha]\!])$ (those species which appear in F_α); and
- *intermediates*$([\![\alpha]\!])$ (the remaining species).

We now abuse the terminology of [11] to define a notion of *copy tolerance*, and use this to state our restrictions on individual reaction encodings.

Definition 13 (Copy tolerance). *A reaction encoding* $[\![\alpha]\!]$ *is* copy tolerant *of a species* x *if* $\{\tau \mid S_\alpha^{init} \vdash_\mathcal{E} \tau\} = \{\tau \mid (S_\alpha^{init} + n \cdot x) \vdash_\mathcal{E} \tau\}$ *for all* $n \in \mathbb{N}$, *where* $\mathcal{E}$ *is the CRN derived from* $[\![\alpha]\!]$ *(extended to include* x *if necessary) and* S_α^{init} *is the initial state of* $[\![\alpha]\!]$.

Definition 14 (Transactional reaction encodings). *Consider a formal reaction* $\alpha = (R_\alpha \rightarrow P_\alpha)$, *encoded as* $[\![\alpha]\!]$ *via the fuel multiset* F_α. *Let* $S_0 = S_\alpha^{init} = \mathcal{M}(R_\alpha) + F_\alpha$ *denote the initial state consisting of just the required reactants and fuels, and let* $\tau = [r_1, \ldots, r_n] \in \text{Traces}(\mathcal{E})$ *be a terminal trace of the encoding starting from* S_0, *where* $r_i = (R_i \rightarrow P_i)$ *for* $i \in \{1, \ldots, n\}$. *Labelling the corresponding sequence of states as* $S_0 \xrightarrow{r_1} S_1 \xrightarrow{r_2} \cdots \xrightarrow{r_{n-1}} S_{n-1} \xrightarrow{r_n} S_n$, *we say that* r_j *is a* commit reaction *if the following criteria are all satisfied:*

1. r_j *is the first irreversible reaction in* τ;
2. *if* $x \in$ *formals*$([\![\alpha]\!])$ *occurs in* $r_1, \ldots, r_{j-1}$ *or* R_j *then* $x \in \mathcal{M}(R_\alpha)$, *and these occurrences are all either reactants of forward steps or products of backward steps;*
3. *if* $x \in$ *formals*$([\![\alpha]\!])$ *occurs in* P_j *or* $r_{j+1}, \ldots, r_n$ *then* $x \in \mathcal{M}(P_\alpha)$, *and these occurrences are all either products of forward steps or reactants of backward steps;*
4. $\mathcal{M}^{-1}(S_{j-1} - R_j) = \varnothing$.

We say that $[\![\alpha]\!]$ *is* transactional *if every terminal trace from* S_0 *has a commit reaction satisfying the above criteria and if every terminal trace visits the same set of states prior to the commit reaction. We also require that* $[\![\alpha]\!]$ *is copy tolerant of all formal and fuel species involved in the encoding, and that the terminal state has the form* $\mathcal{M}(P_\alpha) + L_\alpha$, *where* $[\![\alpha]\!]$ *is copy tolerant of every species in the multiset* L_α *of leftover species.*

In Definition 14, criterion 1 requires that the trace can be partitioned into two disjoint subtraces by the first irreversible reaction. Given the reaction rules from Figure 1, this implies that all reactions before that point must be reversible. Criteria 2 and 3 require that only input formal species can engage in reactions before the commit reaction, and only output formal species can engage in reactions after, and furthermore that the consumption of input formal species before the commit reaction and the production of output formal species after the commit reaction always drives the system forwards. Criterion 4 ensures that all necessary reactants are in fact consumed by the time the commit reaction is reached (this is needed in case the reactants and products have some species in common). The restrictions on copy tolerance ensure that the behaviour of the encoding is identical in the presence of additional copies of fuels or formal species: note that any encoding is copy tolerant of waste species, but that the encoding may not be copy tolerant to certain intermediate species. The restrictions on leftover species in the terminal state delimit those species which may be safely left behind by a reaction encoding that does not fully garbage collect its intermediate species.

Example 1 (Effect of a non-transactional reaction encoding). As a concrete example [12] of what might go wrong when using a non-transactional reaction encoding, consider the following set of formal reactions: $\{x \rightarrow y, y + a \rightarrow y + b\}$, and suppose that our encoding of $x \rightarrow y$ is not transactional because the output y can be released before the first irreversible step in the execution of the encoding. Then, from an initial state corresponding to the formal state $\{x = 1, a = 1\}$ the following sequence of operations is possible:

1. Run the encoding of $x \rightarrow y$ until y is produced, but *without* executing any irreversible steps. This produces a new state corresponding to $\{y = 1, a = 1\}$.
2. Completely execute the encoding of $y + a \rightarrow y + b$, which results in a state corresponding to $\{y = 1, b = 1\}$.
3. Unwind the partial execution of $x \rightarrow y$, which is possible because no irreversible steps have been executed in this encoding. The final state corresponds to the formal state $\{x = 1, b = 1\}$.

Note that the formal state $\{x = 1, b = 1\}$ is not reachable from the initial state $\{x = 1, a = 1\}$ using the above set of formal reactions: our encoding of this CRN is unsound. The specific problem here is that the y produced by the $x \rightarrow y$ reaction is accessible before the encoding has executed an irreversible step to commit to its production. Until there is no way for this product to be reclaimed by the gate that produced it, it is unsound for other reactions to consume it.

We now define *compatible* reaction encodings, in which direct sharing of species between the encodings is only permitted in certain situations.

Definition 15 (Compatible reaction encodings). *We say that two reaction encodings, $[\![\alpha]\!]$ and $[\![\beta]\!]$, are compatible if every shared species in species($[\![\alpha]\!]$) $\cap$ species($[\![\beta]\!]$) appears in the same category (formal, waste, fuel or intermediate) in both encodings, and if both encodings are copy tolerant of every shared species. Furthermore, we require that a species from $[\![\alpha]\!]$ can only interact with a species from $[\![\beta]\!]$ if at least one of those species occurs in species($[\![\alpha]\!]$) $\cap$ species($[\![\beta]\!]$).*

Hence, different reaction encodings may share formal species, waste species and fuel strands. They may also share intermediate species provided that the presence of additional copies of those species do not enable additional reaction pathways in either reaction encoding. In all cases, shared species must appear in the same category in both reaction encodings, and no species may interact with any species from a reaction encoding in which it is not present as a species. We can use the DSD semantics and compiler to check for unwanted interference between reaction encodings. We now state some preliminary lemmas needed to prove our main result.[2]

Lemma 3 (Trace rewriting and reversible reactions). *If $\tau \in$ Traces($\mathcal{C}$) consists entirely of reversible reactions and $\mathbf{S} \vdash_{\mathcal{C}} \tau$ then there exists $\tau' \in$ Traces($\mathcal{C}$) such that $\mathbf{S} \vdash_{\mathcal{C}} \tau{:}\tau' \rightsquigarrow \epsilon$.*

Lemma 4 (Serializability). *Assume that all reaction encodings are transactional and pairwise compatible, and suppose that $\mathbf{S} \vdash_{\mathcal{E}} \tau$, where*

$$\tau = \tau_1{:}[r_1^\alpha]{:}\cdots{:}\tau_{k-1}{:}[r_{k-1}^\alpha]{:}\tau_k{:}[r_{com}^\alpha]{:}\tau_{k+1}{:}[r_{k+1}^\alpha]{:}\cdots{:}\tau_n{:}[r_n^\alpha]{:}\tau_{rest},$$

where r_{com}^α is the first commit reaction in τ, where $\tau_\alpha = [r_1^\alpha, \ldots, r_{k-1}^\alpha, r_{com}^\alpha, r_{k+1}^\alpha, \ldots, r_n^\alpha]$ is an execution of $[\![\alpha]\!]$ and where $\mathbf{S} \vdash_{\mathcal{C}} r_1^\alpha$. Then, there exists τ'_{rest} such that $\mathbf{S} \vdash_{\mathcal{E}} \tau \rightsquigarrow \tau_\alpha{:}\tau'_{rest}$. □

We can now state and prove our main soundness theorem, which is valid for systems composed of reaction encodings that satisfy the criteria in Definition 14 and Definition 15. Since the set of all traces includes incomplete executions of reaction encodings, we require that any trace can be *extended* to produce a serializable execution. In doing so we write $pt(\mathbf{X}, \mathbf{F})$ for the set of non-empty formal traces that are valid from the formal state $\mathbf{X}$ and that can be emulated using the fuel $\mathbf{F}$, i.e., $pt(\mathbf{X}, \mathbf{F}) = \{\tau \in$ Traces($\mathcal{F}$) $\mid \mathbf{X} \vdash_{\mathcal{F}} \tau \wedge reqfuel(\tau) \leq \mathbf{F} \wedge \tau \neq \epsilon\}$.

Theorem 2 (Soundness). *For a formal CRN $\mathcal{F}$ with reactions $\alpha_1, \ldots, \alpha_n$, assume that the corresponding reaction encodings are all transactional and pairwise compatible. Let $\mathbf{X}$ range over multisets of formal species, and let $\mathbf{F}$ range over multisets of fuels such that $\mathbf{F} = \mathbf{F}_{\alpha_{k_1}} + \cdots + \mathbf{F}_{\alpha_{k_j}}$, for $k_1, \ldots, k_j \in \{1, \ldots, n\}$ (i.e., there are no incomplete reaction encodings). Then, for all $\tau \in$ Traces($\mathcal{E}$) and all $\mathbf{X}$ and $\mathbf{F}$ such that $(\mathcal{M}(\mathbf{X}) + \mathbf{F}) \vdash_{\mathcal{E}} \tau$, either:*

- *$pt(\mathbf{X}, \mathbf{F}) = \varnothing$ and there exists τ' such that $(\mathcal{M}(\mathbf{X}) + \mathbf{F}) \vdash_{\mathcal{E}} \tau{:}\tau' \rightsquigarrow \epsilon$; or*
- *$pt(\mathbf{X}, \mathbf{F}) \neq \varnothing$ and there exists τ' such that $(\mathcal{M}(\mathbf{X}) + \mathbf{F}) \vdash_{\mathcal{E}} \tau{:}\tau' \rightsquigarrow \tau_{ser}$, where τ_{ser} is a serial execution of some $\tau_{formal} \in pt(\mathbf{X}, \mathbf{F})$.* □

6 Verification Example

In this section we present an example application of our modular verification strategy to a two-domain strand displacement network [5]. We focus on the two-domain catalyst gate introduced in [6], which implements a reaction of the form $x + y \to x + z$. The DSD code for this gate design is as follows.

---

[2] Proof sketches for Lemma 4 and Theorem 2 are included in the appendices, which can be downloaded from the first author's webpage.

```
(* DSD code for two-domain catalyst gate. *)
(* Use with Infinite DSD semantics. *)

(* Define a global toehold *)
new t

(* Catalyst gate module, x + y -> x + z *)
def C(N,x,y,z) = new a new c
( N * {t^*}[x t^]:[y t^]:[c]:[a t^]:[a]
| N * [x]:[t^ z]:[c]:[t^ y]:[t^ a]{t^*}
| N * <t^ c a>
| N * <z c t^> )

(* Example initial state *)
( C(1,x,y,z) | <t^ x> | <t^ y> )
```

The corresponding initial and terminal states of one such reaction gate implementing the reaction $x + y \to x + z$ are shown in Figure 2. In previous work [6], we used these gates to implement the approximate majority voting circuit of [13], by instantiating the module to implement the four chemical reactions from the approximate majority circuit: (i) $x + y \to y + b$, (ii) $x + y \to x + b$, (iii) $x + b \to x + x$, (iv) $y + b \to y + y$. Thus the catalyst gate module must function correctly both when the two products are different species (for reactions (i) and (ii)) and when they are the same (for reactions (iii) and (iv)). It is not difficult to show that each of the resulting reaction encodings satisfies the correctness criteria from Definition 14, and that they are pairwise compatible (Definition 15). The private domains declared within the scope of the gate definition ensure that each reaction encoding involves unique gate species and fuel strands, and the DSD reaction rules from Figure 1 can be used to verify that there are no interactions between the species of the different reaction encodings.[3] Therefore, by Theorem 1 and Theorem 2 every trace produced by these reactions can be rewritten to produce a serial execution of the four formal reactions above, so we view these gates as a correct encoding of the approximate majority system.

7 Discussion

We have shown that any strand displacement system composed of reaction encodings that meet the criteria from Definition 14 and Definition 15 is correct in the sense that all traces can be rewritten using the rules from Definition 11 into a serialized trace in which each execution of a reaction encoding runs to completion before the next one starts. This notion of correctness is reasonable because reaction gates are intended to encode a single rewriting step in the formal reactions, and if a trace cannot be rewritten in this way there must be a concurrency bug in the reaction encodings that allows them to produce

---

[3] Further details are included in the appendices, which can be downloaded from the first author's webpage.

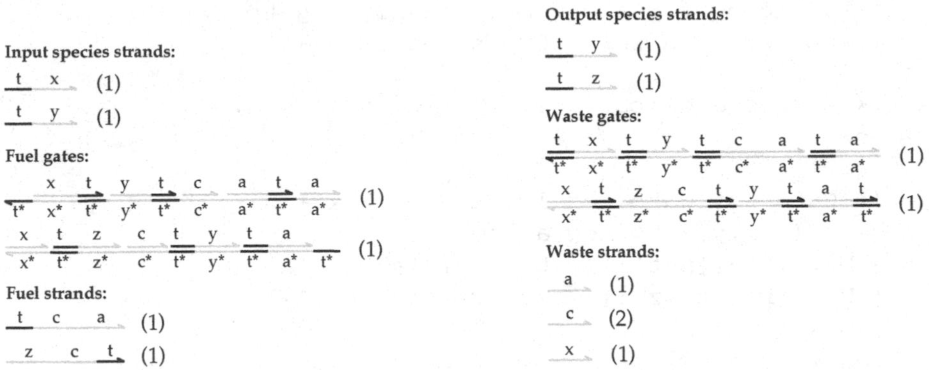

Input species strands:

$\underline{\text{t}\quad\text{x}}$ (1)

$\underline{\text{t}\quad\text{y}}$ (1)

Fuel gates:

$$\frac{\text{x}\quad\text{t}\quad\text{y}\quad\text{t}\quad\text{c}\quad\text{a}\quad\text{t}\quad\text{a}}{\text{t*}\quad\text{x*}\quad\text{t*}\quad\text{y*}\quad\text{t*}\quad\text{c*}\quad\text{a*}\quad\text{t*}\quad\text{a*}} \quad (1)$$

$$\frac{\text{x}\quad\text{t}\quad\text{z}\quad\text{c}\quad\text{t}\quad\text{y}\quad\text{t}\quad\text{a}}{\text{x*}\quad\text{t*}\quad\text{z*}\quad\text{c*}\quad\text{t*}\quad\text{y*}\quad\text{t*}\quad\text{a*}\quad\text{t*}} \quad (1)$$

Fuel strands:

$\underline{\text{t}\quad\text{c}\quad\text{a}}$ (1)

$\underline{\text{z}\quad\text{c}\quad\text{t}}$ (1)

Output species strands:

$\underline{\text{t}\quad\text{y}}$ (1)

$\underline{\text{t}\quad\text{z}}$ (1)

Waste gates:

$$\frac{\text{t}\quad\text{x}\quad\text{t}\quad\text{y}\quad\text{t}\quad\text{c}\quad\text{a}\quad\text{t}\quad\text{a}}{\text{t*}\quad\text{x*}\quad\text{t*}\quad\text{y*}\quad\text{t*}\quad\text{c*}\quad\text{a*}\quad\text{t*}\quad\text{a*}} \quad (1)$$

$$\frac{\text{x}\quad\text{t}\quad\text{z}\quad\text{c}\quad\text{t}\quad\text{y}\quad\text{t}\quad\text{a}\quad\text{t}}{\text{x*}\quad\text{t*}\quad\text{z*}\quad\text{c*}\quad\text{t*}\quad\text{y*}\quad\text{t*}\quad\text{a*}\quad\text{t*}} \quad (1)$$

Waste strands:

$\underline{\quad\text{a}\quad}$ (1)

$\underline{\quad\text{c}\quad}$ (2)

$\underline{\quad\text{x}\quad}$ (1)

Fig. 2. Initial (left) and terminal (right) states for the catalyst gate which encodes the reaction $x+y \rightarrow y+z$.

a trace unrelated to any trace of the underlying formal reactions. Although we used two-domain strand displacement reactions to define our encodings, in principle similar results could be derived for other implementations of chemical reaction networks. It is interesting that the correctness criteria from Definition 14 share much in common with other notions from existing concurrency theory, such as *two-phase locking* [7], in which each transaction has an initial phase of lock acquisition where exclusive access is obtained to the necessary resources, followed by a phase of lock release where those access rights are gradually relinquished. In our case, the first phase consumes the inputs and the second phase produces the outputs.

To our knowledge, this paper is the first modular analysis of chemical reaction network encodings. In Definition 14 and Definition 15 we aimed to allow the maximum possible sharing of species between different reaction encodings without invalidating the soundness result in Theorem 2. However, it is worth noting that certain previously published designs for two-domain strand displacement gates fall foul of our restrictions on the structure of reaction encodings and on the sharing of species between encodings. The gate designs from [5] without irreversible product release do not involve a commit reaction as defined in Definition 14, and therefore the soundness theorem does not hold for these gates. Furthermore, certain combinations of these gates may violate our requirement that shared species must fall into the same category in all reaction encodings. In some gates from [5] it is possible for certain global cosignals to serve as an intermediate in one reaction encoding and as a fuel in another, which could adversely affect the kinetics of the reactions producing that strand as an intermediate if an excess quantity of that strand is supplied as fuel. This subtle point should be addressed in future two-domain gate designs.

Our definition of species encodings works for any scheme in which there is a bijective mapping between the formal species and the DSD species which represent them, such as the two-domain scheme. A straightforward generalization of our representation language to handle *wildcard domains* should allow us to verify gate designs such as those which use three- and four-domain species encodings using history domains [2,3].

In these schemes a single formal species is encoded by a family of related DNA species with similar structure but a different history domain. To extend our results to history domains, it would suffice to prove that each reaction encoding can accept any variant of the input strand, regardless of the history domain. Our other key results, such as serializability and soundness, do not rely on a bijective species encoding and should remain valid. Finally, we have demonstrated that the restrictions we imposed on reaction encodings are *sufficient* to obtain a serializability result. Another important future research direction will be to determine which restrictions are *necessary* to derive such a result.

Prior work on CRN verification [12,14] has focused on analyzing full systems, and we believe that there is a strong connection between our work and the weak bisimulation-based approach of [14]. In other related work, Cardelli and Laneve [15,16] developed a theory of reversible computational structures with a strong relationship to DNA strand displacement reaction gates. However, that work did not take the initial state of a computation into account and was therefore not capable of distinguishing between traces where reactions involving the same species could be safely permuted. Existing work on reachability in CRNs [11], and in particular the notion of *copy tolerance*, is directly related to our work, as is previous work on CRN programmability [17,18].

Our modular approach provides a path to verification of *module definitions*, for example, checking that a definition which maps arbitrary species w, x, y and z to the corresponding reaction encoding $[\![w + x \rightarrow y + z]\!]$ produces a correct reaction encoding for *any* values of w, x, y and z, some of which might in fact represent the same formal species. Note that in Section 6 we did not verify the module definition directly, but rather a number of specific instantiations of it. Expressing our correctness criteria in a temporal logic would allow module definitions to be checked automatically using a model checker, in order to cover all possible input patterns.

Finally, we note that our formalism and proofs do not take account of reaction rates. Expressing correctness in terms of reachability is both important and natural from a computer science perspective. However, unfavourable kinetics might cause gates that satisfy our reachability criteria to function poorly in practice, as discussed above. Furthermore, certain gate designs that fail to satisfy the criteria might function acceptably in practice due to favourable kinetics, as exemplified by the "wisdom of crowds" example [6,5]. Proving soundness of gate designs is already challenging without considering reaction rates, and indeed it is not clear how such a correctness result would be formulated in a modular setting when considering reaction rates. Previous work [2] presented similar proofs for a particular encoding of chemical reaction networks using DNA strand displacement, and future extensions of our work may enable such results to be proved for arbitrary reaction encodings in a modular way. For instance, it may be possible to relate the expected time to fully execute a reaction encoding to the rate of the corresponding formal reaction, either by solving the corresponding continuous-time Markov chain analytically or by using a probabilistic model checker such as PRISM [19]. However, such efforts would be complicated by the fact that the output species from a reaction encoding are typically released gradually, some time before the final irreversible reaction that concludes the execution of the encoding. Hence, it is not obvious which point in time should be considered as the end of the execution of the encoding for the purposes of proving results about the kinetics.

Acknowledgments. This material is based upon work supported by the National Science Foundation under grants 1027877 and 1028238. M.R.L. gratefully acknowledges support from the New Mexico Cancer Nanoscience and Microsystems Training Center.

References

1. Zhang, D.Y., Seelig, G.: Dynamic DNA nanotechnology using strand-displacement reactions. Nat. Chem. 3(2), 103–113 (2011)
2. Soloveichik, D., Seelig, G., Winfree, E.: DNA as a universal substrate for chemical kinetics. PNAS 107(12), 5393–5398 (2010)
3. Cardelli, L.: Strand algebras for DNA computing. Nat. Comput. 10(1), 407–428 (2010)
4. Lakin, M.R., Youssef, S., Cardelli, L., Phillips, A.: Abstractions for DNA circuit design. JRS Interface 9(68), 470–486 (2012)
5. Cardelli, L.: Two-domain DNA strand displacement. Math. Structures Comput. Sci. 23, 247–271 (2013)
6. Lakin, M.R., Parker, D., Cardelli, L., Kwiatkowska, M., Phillips, A.: Design and analysis of DNA strand displacement devices using probabilistic model checking. JRS Interface 9(72), 1470–1485 (2012)
7. Papadimitriou, C.H.: The serializability of concurrent database updates. Journal of the ACM 26(4), 631–653 (1979)
8. Chen, H.-L., Doty, D., Soloveichik, D.: Deterministic function computation with chemical reaction networks. In: Stefanovic, D., Turberfield, A. (eds.) DNA18. LNCS, vol. 7433, pp. 25–42. Springer, Heidelberg (2012)
9. Phillips, A., Cardelli, L.: A programming language for composable DNA circuits. JRS Interface 6(Suppl. 4), S419–S436 (2009)
10. Lakin, M.R., Youssef, S., Polo, F., Emmott, S., Phillips, A.: Visual DSD: A design and analysis tool for DNA strand displacement systems. Bioinformatics 27(22), 3211–3213 (2011)
11. Condon, A., Kirkpatrick, B., Maňuch, J.: Reachability bounds for chemical reaction networks and strand displacement systems. In: Stefanovic, D., Turberfield, A. (eds.) DNA18. LNCS, vol. 7433, pp. 43–57. Springer, Heidelberg (2012)
12. Shin, S.W.: Compiling and verifying DNA-based chemical reaction network implementations. Master's thesis, California Institute of Technology (2012)
13. Angluin, D., Aspnes, J., Eisenstat, D.: A simple population protocol for fast robust approximate majority. Dist. Comput. 21(2), 87–102 (2008)
14. Dong, Q.: A bisimulation approach to verification of molecular implementations of formal chemical reaction networks. Master's thesis, Stony Brook University (2012)
15. Cardelli, L., Laneve, C.: Reversible structures. In: Fages, F. (ed.) CMSB 2011 Proceedings, pp. 131–140. ACM (2011)
16. Cardelli, L., Laneve, C.: Reversibility in massive concurrent systems. Sci. Ann. Comp. Sci. 21(2), 175–198 (2011)
17. Soloveichik, D., Cook, M., Winfree, E., Bruck, J.: Computation with finite stochastic chemical reaction networks. Nat. Comput. 7, 615–633 (2008)
18. Cook, M., Soloveichik, D., Winfree, E., Bruck, J.: Programmability of chemical reaction networks. In: Condon, A., Harel, D., Kok, J.N., Salomaa, A., Winfree, E. (eds.) Algorithmic Bioprocesses, pp. 543–584. Springer (2009)
19. Kwiatkowska, M., Norman, G., Parker, D.: PRISM 4.0: Verification of probabilistic real-time systems. In: Gopalakrishnan, G., Qadeer, S. (eds.) CAV 2011. LNCS, vol. 6806, pp. 585–591. Springer, Heidelberg (2011)

Iterative Self-assembly with Dynamic Strength Transformation and Temperature Control

Dandan Mo and Darko Stefanovic

Department of Computer Science, University of New Mexico,
MSC01 1130 1 University of New Mexico, Albuquerque, NM 87131-0001, U.S.A.
{mdd,darko}@cs.unm.edu

Abstract. We propose an iterative approach to constructing regular shapes by self-assembly. Unlike previous approaches, which construct a shape in one go, our approach constructs a final shape by alternating the steps of assembling and disassembling, increasing the size of the shape iteratively. This approach is embedded into an extended hexagonal tile assembly system, with dynamic *strength transformation* and *temperature control*. We present the construction of equilateral triangles as an example and prove the uniqueness of the final shape. The tile complexity of this approach is $O(1)$.

Keywords: Algorithmic self-assembly, hexagonal tiles, strength transformation, temperature control.

1 Introduction

The *tile assembly system* (TAS) contains a self-assembly process in which small tiles autonomously attach to a seed, assembling into a larger and more complex shape. TAS dates back to the late 1990s. In 1998, Winfree proposed a mathematical model of DNA self-assembly [1] with the operations of *Annealing*, *Ligation*, and *Denaturation*. The same year, Winfree et al. presented a tile assembly model [2] which connected tiling theory with structural DNA nanotechnology, using the double-crossover molecules (DX) to implement the model. In 2000, Rothemund and Winfree proposed the *abstract tile assembly model* (aTAM) [3], in which the tiles are squares with sticky ends on each side. Adleman et al. extended this model with stochastic *time complexity* [4].

Later works concerning the aTAM or TAS have been mainly extensions aimed at improving the complexity or exploring more final shapes that can be self-assembled. Instead of the original square tiles, triangular, hexagonal [5,6], and string tiles [7] can also implement the TAS. Aggarwal et al. extended the standard TAS by allowing the assembly of the super tiles [8]. Demaine et al. proposed a staged self-assembly model, where the tiles can be added in sequence [9]. The tile complexity problem usually studies the cost of building an $N \times N$ square in terms of the number of tile types required. Compared with the previous result N^2 [3] at temperature[1] 1, geometric tiles [10] use only $\Theta(\sqrt{\log N})$ tile types. An extended model with a mechanism of temperature programming [11,12] uses only $O(1)$ tile types. Besides an $N \times N$ square, other regular

---

[1] The temperature is a threshold value that determines if a shape is stable.

D. Soloveichik and B. Yurke (Eds.): DNA 2013, LNCS 8141, pp. 147–159, 2013.

shapes can be self-assembled using the TAS. Cook et al. showed that a Sierpinski triangle can be formed by 7 different square tiles [13]. Kari et al. presented a triangular tile self-assembly system [14] where an equilateral triangle of size N can be formed using N^2 tile types at temperature 1, and $2N - 1$ tile types at temperature 2. Woods et al. proposed an active self-assembly model [15] to construct algorithmic shapes and patterns in polylogarithmic time.

Here we propose an iterative approach to self-assembly, with a tile complexity of $O(1)$. The approach extends a tile assembly system with dynamic *strength transformation* and *temperature control*. Unlike previous TAS models, where the final stable shape is formed in one go, our approach constructs regular shapes of increasing size in each iteration in a geometric progression. This iterative process consists of assembling an n-size shape from an n-size seed cluster[2] and disassembling it into two pieces which then form a new seed cluster with a larger size for the next iteration. In other words, our approach alternates the steps of *assembling* and *disassembling*. We demonstrate how this approach works on the example of constructing equilateral triangles; the rest of the paper treats a specific tile assembly system for triangles.

In the following, we first give an overview of our iterative approach in Section 2. In Section 3, a formal definition of the extended hexagonal TAS is given. In Section 4, we present a concrete example of enlarging an equilateral triangle from size four to size six[3]. In Section 5, we prove that an equilateral triangle is the only final stable shape that can be self-assembled within this system.

2   Overview of the Approach

Our iterative approach repeats two steps: assembling and disassembling. In the assembling step, we start from a seed cluster, which is one of the three sides of the triangle when an iteration ends. The free individual tiles present in solution attach to this seed cluster, updating the shape until an equilateral triangle is completed. The assembling step takes place under a constant temperature τ_1, which is also the threshold value to determine if a shape is stable[4]. The disassembling step begins when we raise the temperature to τ_2, which will make the τ_1-stable shape become unstable under τ_2, and lead to disassembly. The original shape then disassembles into two smaller shapes that are τ_2-stable. We then lower the temperature down to τ_3 to make the two shapes form a new seed cluster with larger size. When we change the temperature from τ_3 back to τ_1, the next iteration starts.

The assembling and disassembling step are enabled by two new features added to the hexagonal TAS. *Strength transformation* changes some of the bond strengths of the τ_1-stable shape and two strand strengths, in preparation for the disassembling step. *Temperature control* implements the switch between the two steps. These two features are supported by the **Rule** set and the **Operation** set in the extended hexagonal TAS, which we will discuss in the next section.

---

[2] A seed cluster of size n is a line consisting of n hexagonal tiles.

[3] The size of an equilateral triangle is denoted as its edge length.

[4] If a shape does not disassemble under temperature τ, it is τ-stable.

3 Formal Definition of the Extended Hexagonal TAS

In this section, we give a formal expression of the extended hexagonal TAS, with detailed explanations of each term. We then introduce the set of tile types in this system and illustrate the whole procedure of our iterative approach. Finally, we consider plausible chemical implementations of some operations included in the *strength transformation*.

3.1 A Formal Expression

The extended hexagonal TAS is formally a tuple $M = \{\Sigma, S, B, \Omega, U, R, O\}$, where Σ is a temperature set, S is a strand set, B is a bond set, Ω is a tile set, U is a stable shape set, R is a rule set, and O is an operation set.

Temperature Set. $\Sigma = \{\tau | \tau \in \mathbb{Z}_{>0}\}$ is the set of the temperatures.

In our example of constructing an equilateral triangle, there are three threshold temperatures in this system. $\Sigma = \{\tau_1 = 3, \tau_2 = 7, \tau_3 = 5\}$.

Strand Set And Bond Set. $S = \{x | x \in A\} \cup \{\bar{x} | x \in A\}$ is the set of DNA strand types, where A denotes a symbol set. Two complementary strands x and $\bar{x}$ can form a bond with the corresponding strength. We use $bond(x, \bar{x} \parallel w)$ to represent a bond and its sticking strength w, where $w \in \mathbb{Z}_{>0}$. B is the set of bonds. Thus, for all $b \in B$, we have $b = bond(x, \bar{x} \parallel w)$ such that $x, \bar{x} \in S$. Here, w is the bond strength as well as the strength of strand x and $\bar{x}$.

In our example, $S = \{a, y, \gamma, Y\} \cup \{\bar{a}, \bar{y}, \bar{\gamma}, \bar{Y}\}$. Table 1 shows all the strands we use. For example, $a, \bar{a}$ form a bond with strength 1.

Tile Set. Ω is the set of tiles. Each tile $t = (s_1, s_2, s_3 ..., s_6)$ is abstracted as a hexagon, each side represents a strand. An example is shown in Figure 1. In our graphical representation, if a side does not have any strand, we use the dotted line to represent it. A pair of complementary strands, e.g., a and $\bar{a}$, are represented by a single solid line and a double solid line respectively. The number labeled next to the strand indicates the strand strength. The sequence of $s_i (1 \leq i \leq 6) \in S \cup \{\perp\}$ represents each side of the hexagon from the top right clockwise to the top left. A side without any strand is denoted as $\perp$. Tiles are allowed to rotate, but not to flip. Therefore, if there are two tiles $t_1 = (s_1, s_2 ..., s_6)$, $t_2 = (s_1', s_2' ..., s_6')$ and if $\exists i, j \ (1 \leq i, j \leq 6 \text{ and } j = 7 - i)$ such that $(s_i', s_{i+1}' ..., s_6') = (s_1, s_2 ..., s_j)$ and $(s_1', s_2' ..., s_{i-1}') = (s_{j+1}, s_{j+2} ..., s_6)$, we say that t_1 and t_2 are the same tile. We assume that the strands of the tiles are fixed in their positions, thus the angles between each two are fixed as well.

Stable Shape Set. U is the set of stable shapes. 1. $\forall t \in \Omega, t \in U$; 2. each shape $T \in U$ is either $T = \{t\} \cup T'$ or $T = T_1 \cup T_2$, where $t \in \Omega$ and T', T_1, T_2 are also stable shapes. That is, a stable shape is a single tile or a set of tiles that are grouped together because of the bonds between them. A stable shape can be formed in two ways—either a single tile

Table 1. Strand strength

Strand type x	$a, \bar{a}$	$y, \bar{y}$	$\gamma, \bar{\gamma}$	$Y, \bar{Y}$
Strand strength	1	2	5	7

Fig. 1. This figure shows an example of a hexagon tile: $t = (y, \bar{a}, a, \bar{y}, \perp, \perp)$. The numbers labeled next to the lines indicates the strengths of the strands.

attaches to an existing stable shape or two stable shapes attach to each other. Both ways are based on the premise that the sum of the interacting strengths equals or exceeds the current temperature, which is expressed as the "assembling" rule that will be discussed shortly.

Rule Set. $R = \{assembling, stability - checking, disassembling\}$ is a set of rules that guide this system, shown in Figure 2.

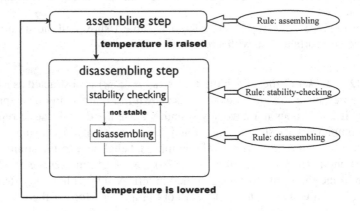

Fig. 2. How the rules guide the steps of assembling and disassembling

If the temperature is not changed, the assembling rule is applied. Whenever the temperature is raised, the stability-checking rule and the disassembling rule are applied to execute the disassembling step. We use the stability-checking rule to check the stability of the shape. If the shape is stable, there is no change to the shape, otherwise the disassembling rule is applied to break the shape. Whenever the temperature is lowered, the assembling rule is applied.

Rule 1: Assembling. The assembling rule guides the procedure of shape formation. For $T = \{t\} \cup T'$ under temperature τ, where T' is an existing stable shape, the interaction strength sum E between t and T' must be $E \geq \tau$, otherwise T will not be stable. Similarly, for $T = T_1 \cup T_2$ where T_1, T_2 are stable, the interaction strength sum E between T_1 and T_2 must be $E \geq \tau$ as well, otherwise shape T will be disassembled into the two smaller but stable shapes T_1 and T_2. Under the assembling rule, the formed shape is guaranteed to be stable.

Rule 2: Stability-Checking. The stability-checking rule is applied to check the stability of a shape after the temperature is raised. There are two levels of checking. Suppose that we have a shape T and the temperature is raised to τ. We first check if each tile $t \in T$ is stable in the shape, then check if every two sub-shapes T_1, T_2 are stably attached to each other($T_1 \cup T_2 = T$) if the first-level stability is not achieved. In the first level, for every tile $t \in T$, we compare the sum of the interaction strength between t and $T - t$ with τ_2. If the sum $< \tau$, T is not stable. Otherwise, we proceed to the second-level checking. In the second level, if there exists a cut that splits T into two sub-shapes T_1, T_2 and the sum of interaction strengths between T_1 and T_2 is less than τ, T is not stable. Only when both levels of checking succeed, can we say that shape T is stable.

Rule 3: Disassembling. The disassembling rule is applied when the shape is not stable. It works together with the stability-checking rule to disassemble the shape into smaller stable shapes. There are two levels of disassembling. The first level makes certain single tiles fall off from the shape; the second level splits the shape into two new shapes. In the first-level disassembling, we execute the first-level stability checking to find the tiles whose interaction strengths are the weakest among those tiles with interaction strength less than the current temperature, and remove these tiles from the shape. In the second-level disassembling, we execute the second-level stability checking to find the cut that needs the minimum strengths, and split the shape into two new shapes. The first-level and the second-level disassembling cannot happen at the same time, the one that needs minimum strengths takes place first. If the strengths needed are the same, the first-level disassembling comes first. The stability-checking rule is applied to the new shapes after either level disassembling is executed. If the shapes are not stable, the disassembling rule is applied to the current shapes, otherwise the disassembling step ends.

Operation Set. $O = O_1 \cup O_2 \cup \{End\}$ is a set of operations allowed in the system. Set O_1 includes the operations related to the strength transformation. These operations need the help of some specific restriction enzymes, ligase and auxiliary tiles. Operation $Op_{i \to j} \in O_1(i, j \in \mathbb{Z}_{>0})$ changes the strength from i to j. Set O_2 includes the operations related to the temperature control, operation $\tau_i \to \tau_j \in O_2(\tau_i, \tau_j \in \Sigma)$ changes the temperature from τ_i to τ_j. The last set contains only one operation. When End is executed, the construction procedure ends. These operations are executed when the current shapes are all stable, in other words, during an assembling step or a disassembling step, the experimentalist should not execute any operation.

In our example, $O = \{Op_{5 \to 2}, ligation, Op_{2 \to 5}\} \cup \{\tau_1 \to \tau_2, \tau_2 \to \tau_3, \tau_3 \to \tau_1\} \cup \{End\}$. $Op_{5 \to 2}$ changes a bond from strength 5 to strength 2(with Enzyme I, Enzyme II), $Op_{2 \to 5}$

changes a strand from strength 2 to strength 5(with Enzyme III and the $Op_{2\rightarrow5}$ helper). The operation of *ligation* needs the help of a ligase, changing a bond from strength 2 to strength 7. Enzyme I, Enzyme II and Enzyme III are restriction enzymes, which will be discussed with details in Section 3.4. The $Op_{2\rightarrow5}$ helper is explained in Section 3.2.

3.2   Tile Types

In the system of constructing an equilateral triangle, there are four categories of tiles: initial component tiles, free tiles, functional tiles, and variant tiles. The first three are directly made manually; the variant tiles result from reactions with enzymes. Initial component tiles form a seed cluster of size four; it is the first stable shape existing in the solution before any reaction begins. Free tiles remain in the solution all the time. Functional tiles come in two flavors, one ($Op_{2\rightarrow5}$ helper) is included in the operation $Op_{2\rightarrow5}$, while the other (ending tile) will end the whole procedure with an equilateral triangle. The functional tiles are added to the solution by the experimentalist at certain time points. Variant tiles only appear when one of the operations in $\{Op_{5\rightarrow2}, ligation, Op_{2\rightarrow5}\}$ is executed. The initial component tiles, free tiles and functional tiles are shown in Figure 3. We will explain the transformations to the variant tiles in Section 3.4, together with the implementations of some operations. In Figure 3, y_{adp} and $\bar{y}_{adp}$ are two special strands on $Op_{2\rightarrow5}$ helper tile. Strands y_{adp} and $\bar{y}$ can form a special bond, and likewise the strands $\bar{y}_{adp}$ and y.

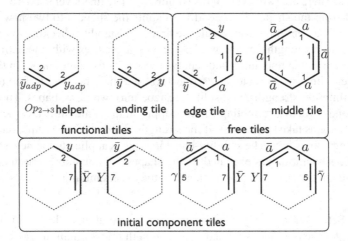

Fig. 3. Tile types is shown in this figure. The dotted lines represent "no strand". Each strand is represented by a single or double solid line with a name label and a strength label.

3.3   The Procedure of the Iterative Approach

The assembling step and the disassembling step together constitute one iteration of the construction. There are two parameters associated with each iteration. One is the seed cluster, which is one edge of the triangle at the beginning of each iteration; the other is the current temperature τ. In each iteration, the assembling step starts with $\tau = 3$ and a seed cluster, and ends with an equilateral triangle with the last piece left; the disassembling step starts with $\tau = 7$ and ends with two symmetric shapes that will form the new seed cluster at $\tau = 5$. When $\tau = 3$, the next iteration starts. Table 2 summarizes the actions during an iteration. The "Action" column is the operation executed at each step, the "State" column expresses the current condition of the shape and temperature. Notice that at step 1 we have two branches — choosing the *End* operation will end the iteration with an equilateral triangle of size l, otherwise we proceed to the next step. At step 2, bond($y, \bar{y}\|2$) is transformed to bond($Y, \bar{Y}\|7$). At step 3, bond($\gamma, \bar{\gamma}\|5$) is broken into strands y and $\bar{y}$ that will form bond($y, \bar{y}\|2$) later.

Table 2. The procedure of an iteration and how to stop the iteration

Step No.	Action	State
0		seed cluster of size l, $\tau = 3$
1		an equilateral triangle with the last piece left
	End	iteration ends with an equilateral triangle of size l
2	*ligation*	bond($y, \bar{y}\|2$) $\rightarrow$ bond($Y, \bar{Y}\|7$)
3	$Op_{5\rightarrow2}$	bond($\gamma, \bar{\gamma}\|5$) $\rightarrow$ y and $\bar{y}$ $\rightarrow$ bond($y, \bar{y}\|2$)
4	$Op_{2\rightarrow5}$	$y \rightarrow \gamma$, $\bar{y} \rightarrow \bar{\gamma}$
5	$\tau_1 \rightarrow \tau_2$	two symmetric shapes, $\tau = 7$
6	$\tau_2 \rightarrow \tau_3$	new seed cluster of size $l = 2l - 2$ is formed, $\tau = 5$
7	$\tau_3 \rightarrow \tau_1$	go to step 0

3.4 Proposed Implementation of the Operations

Figure 4 and Figure 5 show some plausible implementations of operations $Op_{5\rightarrow2}$, $Op_{2\rightarrow5}$, and *ligation*. For each implementation, we provide a high-level picture on the top to show how that operation changes the tiles. Tiles to the right of the arrow are the variant tiles. Below each high-level picture are the details concerning the strands.

In Figure 4, the left column shows the steps of transforming strand $\bar{y}$ into strand $\bar{\gamma}$. The operation orders are important. Enzyme III must be added after the tile $Op_{2\rightarrow5}$ helper attaches to the shape, otherwise it might be difficult to tell if $\bar{\gamma}$ is formed. Enzyme III is a restriction enzyme which is assumed to recognize the specific sequences on B-B' part. When the helper tile attaches to the shape, by observing the shape, we are sure that the intermediate product shown in (2) is formed, which will be transformed into $\bar{\gamma}$ after the cutting in (3). If we add Enzyme III before the helper tile fills in its position, the sequence 2 and 5 might be cut off from the helper tile so that the helper tile cannot attach to the shape. In this case, whether the intermediate product is formed is hard to know, so it is hard to know if strand $\bar{\gamma}$ is formed. The right column follows the similar steps, transforming strand y into strand γ.

In Figure 5, the left part is the implementation of operation $Op_{5\rightarrow2}$, the right part is the implementation of operation *ligation*. In the implementation of $Op_{5\rightarrow2}$, two restriction enzymes are used to cut the sequences at two different positions, which is shown in (2). After the cutting, bond($\gamma, \bar{\gamma}\|5$) is broken into two strands ($y, \bar{y}$) and sequence

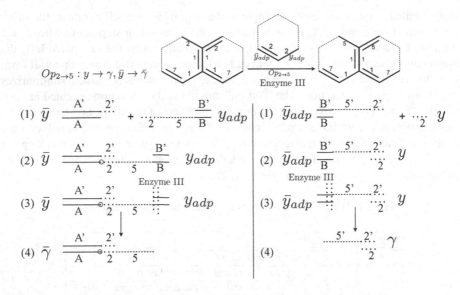

$$Op_{2\to5} : y \to \gamma, \bar{y} \to \bar{\gamma}$$

Fig. 4. This figure shows the plausible implementation of operation $Op_{2\to5}$. Letters A, B and numbers 2, 5 indicate different sequences. A', B', 2' and 5' are the corresponding complementary sequences. There are two sides for each strand, the side with the strand name is fixed on the tile, the other side is free. For example, in (1) of the left column, the left ends of A' and A are fixed, the right end of 2' is free. The left column shows the steps of transforming strand $\bar{y}$ into strand $\bar{\gamma}$. The 2' end on $\bar{y}$ sticks to the 2 end on $\bar{y}_{adp}$, forming (2). Enzyme III is a restriction enzyme which is assumed to recognize the specific sequences on B-B' part. It can cut (2) at the position indicated by the two vertical dotted lines, transforming $\bar{y}$ into $\bar{\gamma}$, which is shown in (3) and (4). The small circle between the sequence A and 2 means no connection. The right column follows the similar steps, transforming strand y into strand γ.

fragments (sequence 2, 5 and sequence 5', 2). Strand y and $\bar{y}$ form a new bond later, which is shown in (3). The sequence fragments leave the tiles. In the implementation of *ligation*, ligase is used to connect the sequence A and 2, transforming bond$(y, \bar{y}\|2)$ into bond$(Y, \bar{Y}\|7)$.

4   Constructing an Equilateral Triangle from Size 4 to 6

Figure 6 and Figure 7 show the complete procedure of constructing an equilateral triangle of size 6 from one of size 4. The number labeled next to the hexagon side indicates the strength of the strand that side represents. The strands with strength 1 are not labeled for convenience in the figures. The initial state is a seed cluster of size 4 and temperature $\tau = \tau_1 = 3$. While the temperature is not changed, the edge tiles and middle tiles in solution attach to the seed cluster one by one, constructing a stable shape shown in state 1. The operation *ligation* transforms the bonds of strength 2 into the bonds of strength 7, updating the shape to state 2. We then execute the $Op_{5\to2}$ operation, which needs the help of Enzyme I and Enzyme II. After that, the shape is updated to state 3, where

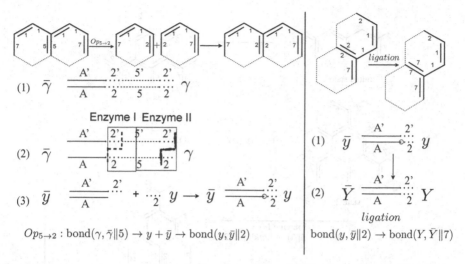

(1) $\bar{\gamma}$ $\dfrac{A' \quad 2' \quad 5' \quad 2'}{A \quad 2 \quad 5 \quad 2}$ γ

Enzyme I Enzyme II

(2) $\bar{\gamma}$ $\dfrac{A' \quad 2' \quad 5' \quad 2'}{A \quad 2 \quad 5 \quad 2}$ γ

(3) $\bar{y}$ $\dfrac{A' \quad 2'}{A}$ $+$ $\cdots_2\, y \rightarrow \bar{y}$ $\dfrac{A' \quad 2'}{A} \circ_2\, y$

$Op_{5\rightarrow 2} : \mathrm{bond}(\gamma, \bar{\gamma}\|5) \rightarrow y + \bar{y} \rightarrow \mathrm{bond}(y, \bar{y}\|2)$

(1) $\bar{y}$ $\dfrac{A' \quad 2'}{A} \circ_2\, y$

(2) $\bar{Y}$ $\dfrac{A' \quad 2'}{A} _2\, Y$

ligation

$\mathrm{bond}(y, \bar{y}\|2) \rightarrow \mathrm{bond}(Y, \bar{Y}\|7)$

Fig. 5. This figure shows the plausible implementations of operation $Op_{5\rightarrow 2}$ and *ligation*. **(a)** The operation $Op_{5\rightarrow 2}$ is implemented by cutting twice. (1) shows the bond formed by γ and $\bar{\gamma}$. In (2), there are two restriction enzymes. We assume that Enzyme I recognizes the sequences within the left square, and it can cut the bond as indicated by the dashed line. We assume that Enzyme II recognizes the sequences within the right square, and it can cut the bond as indicated by the thick line. After the cutting, what are left on the tiles are the strands $\bar{y}$ and y, which can form a new bond$(y, \bar{y}\|2)$ later. **(b)** In the part of *ligation*, the small circle between two sequences means no connection. In (1), the bond formed by y and $\bar{y}$ has strength 2. This strength is proportional to the length of their interacting part, which is 2-2' here. After ligation, parts A and 2 are connected, which means that the interacting part between the two strands is A-A' plus 2-2' now. Since this is longer, the strength of the bond is stronger. We define the new bond formed by A-A' and 2-2' has strength 7, and use bond$(Y, \bar{Y}\|7)$ to represent it.

the bond of strength 5 is transformed into a bond of strength 2. At that point, operation $Op_{2\rightarrow 5}$ is executed. This operation has two phases: in phase (a), the helper tile fills in the top position of the shape, forming two special bonds (the bond consisting of y_{adp}, $\bar{y}$ and the bond consisting of $\bar{y}_{adp}, y$); in phase (b), two restriction enzymes are added into the solution. They find the two special bonds, cut them at a specific position, which transforms y (single solid line with 2) into γ (single solid line with 5) and $\bar{y}$ (double solid line with 2) into $\bar{\gamma}$ (double solid line with 5), updating the shape to state 4. When we raise the temperature from τ_1 to τ_2, the disassembling step begins. The middle tile falls off from the shape first, since its total interaction strengths to the shape sum to 6, less than the current temperature $\tau_2 = 7$. The shape is then disassembled from the middle into two symmetric shapes as shown in state 5, since cutting the shape from the middle only needs strength 3, which is less than τ_2. When we change the temperature from τ_2 to τ_3, two free ends labeled as 5 can form a bond of strength 5, which constructs a new seed cluster as shown in state 6. Finally, we change the temperature to τ_1 again to start the next iteration. In Figure 7, the second iteration starts with a seed cluster of size 6,

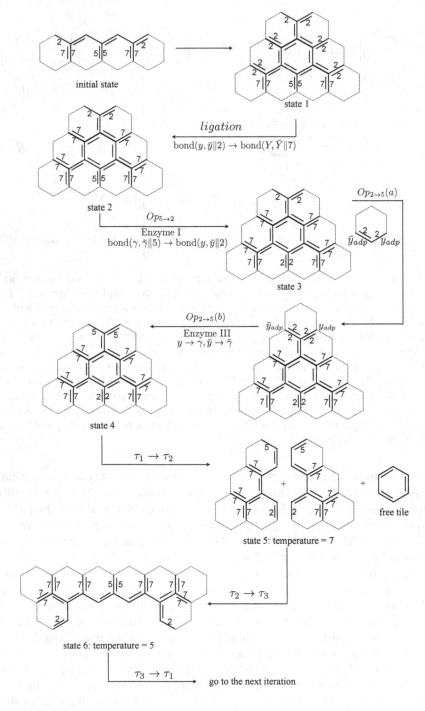

Fig. 6. Constructing an equilateral triangle of side 6 from 4: the first iteration

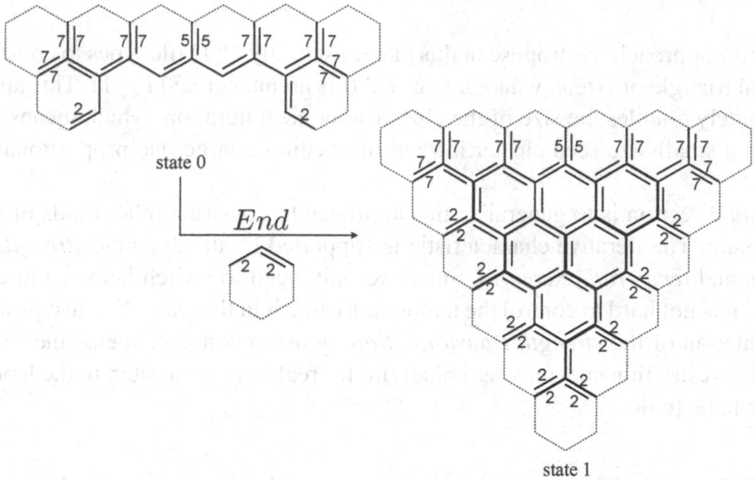

Fig. 7. Constructing an equilateral triangle of side 6 from 4: the second iteration

which is shown in state 0. Putting the ending tile into the solution ends the construction with an equilateral triangle as shown in state 1. The stable final shape is an equilateral triangle of side 6.

5   Equilateral Triangle Is the Only Stable Final Shape

To prove that within this extended TAS the equilateral triangle is the only stable final shape, we need to prove (1) during the assembling step, no other shape except for an equilateral triangle is formed; (2) after the disassembling, there are only two shapes in solution; (3) at temperature 5, the two shapes formed after the disassembling step will bond together to form the new seed cluster for the next iteration.

(1) During the assembling step, $\tau = 3$, no two free tiles can bond together, since the strands on them have strength of 2 at most. There is only one stable shape in solution, which is the seed cluster, so the only way that a new shape is formed is that one free tile attaches to the seed cluster, updating the shape, then comes another free tile. For the edge tile, it can only stick to the leftmost position or the rightmost position of the seed cluster. For the middle tile, it can only stick to the seed cluster with the help of one of its neighbors. That is, the assembling procedure is level-by-level and from-edge-to-center. In addition to the angle design, the only stable shape formed is the equilateral triangle.

(2) When $\tau = 7$, the middle tiles in the one-piece-left triangle start to fall off first since they have the smallest sum of interaction strengths. After that, two bonds $b_1 = bond(a, \bar{a} \| 1)$, $b_2 = bond(y, y \| 2)$ are broken, which splits the current shape into two stable shapes.

(3) When the temperature $\tau = 5$, the free strands that can stick together are γ and $\bar{\gamma}$. Because the two shapes each have one of them, $bond(\gamma, \bar{\gamma} \| 5)$ will be formed, which groups the two shapes together to form a new seed cluster.

6   Conclusion and Future Work

The iterative approach we propose in this paper uses only $O(1)$ tile types to construct an equilateral triangle of size n, where $n = 2^i + 2$ (i is an integer and $i \geq 1$). This approach approximately doubles the size of the cluster after each iteration, which means we can start from a small-size seed cluster instead of needing a large one proportional to the final size.

A future direction is to generalize this approach to construct other kinds of shapes, e.g., a square. The iterative characteristic is supported by the dynamic *strength transformation* and *temperature control*. Since we only need to switch between three temperatures, it is not hard to control the temperature in lab in this case. We also propose an implementation of the *strength transformation* by using conceptual enzymes, a ligase and specific restriction enzymes, as a blueprint for realizing the system in the laboratory which we hope to do.

Acknowledgments. This material is based upon work supported by the National Science Foundation under grants 1027877 and 1028238.

References

1. Winfree, E.: Algorithmic Self-Assembly of DNA (1998)
2. Winfree, E., Liu, F., Wenzler, L.A., Seeman, N.C.: Design and self-assembly of two-dimensional DNA crystals. Nature 394(6693), 539–544 (1998)
3. Rothemund, P.W.K., Winfree, E.: The program-size complexity of self-assembled squares (extended abstract). In: Proceedings of the Thirty-Second Annual ACM Symposium on Theory of Computing, STOC 2000, pp. 459–468. ACM, New York (2000)
4. Adleman, L., Cheng, Q., Goel, A., Huang, M.D.: Running time and program size for self-assembled squares. In: Proceedings of the Thirty-Third Annual ACM Symposium on Theory of Computing, STOC 2001, pp. 740–748 (2001)
5. Chelyapov, N., Brun, Y., Gopalkrishnan, M., Reishus, D., Shaw, B., Adleman, L.: DNA Triangles and Self-Assembled Hexagonal Tilings. Journal of the American Chemical Society 126(43), 13924–13925 (2004)
6. Kari, L., Seki, S., Xu, Z.: Triangular and hexagonal tile self-assembly systems. In: Dinneen, M.J., Khoussainov, B., Nies, A. (eds.) WTCS 2012 (Calude Festschrift). LNCS, vol. 7160, pp. 357–375. Springer, Heidelberg (2012)
7. Winfree, E., Eng, T., Rozenberg, G.: String tile models for DNA computing by self-assembly. In: Condon, A., Rozenberg, G. (eds.) DNA 2000. LNCS, vol. 2054, pp. 63–88. Springer, Heidelberg (2001)
8. Aggarwal, G., Cheng, Q., Goldwasser, M., Kao, M., de Espanes, P., Schweller, R.: Complexities for generalized models of self-assembly. SIAM Journal on Computing 34(6), 1493–1515 (2005)
9. Demaine, E., Demaine, M., Fekete, S., Ishaque, M., Rafalin, E., Schweller, R., Souvaine, D.: Staged self-assembly: nanomanufacture of arbitrary shapes with o(1) glues. Natural Computing 7(3), 347–370 (2008)
10. Fu, B., Patitz, M., Schweller, R., Sheline, R.: Self-assembly with geometric tiles. In: Czumaj, A., Mehlhorn, K., Pitts, A., Wattenhofer, R. (eds.) ICALP 2012, Part I. LNCS, vol. 7391, pp. 714–725. Springer, Heidelberg (2012)

11. Kao, M.Y., Schweller, R.: Reducing tile complexity for self-assembly through temperature programming. In: Proceedings of the Seventeenth Annual ACM-SIAM Symposium on Discrete Algorithm, SODA 2006, pp. 571–580 (2006)
12. Summers, S.: Reducing tile complexity for the self-assembly of scaled shapes through temperature programming. Algorithmica 63(1-2), 117–136 (2012)
13. Cook, M., Rothemund, P., Winfree, E.: Self-assembled circuit patterns. In: Chen, J., Reif, J.H. (eds.) DNA9. LNCS, vol. 2943, pp. 91–107. Springer, Heidelberg (2004)
14. Kari, L., Seki, S., Xu, Z.: Triangular tile self-assembly systems. In: Sakakibara, Y., Mi, Y. (eds.) DNA 16. LNCS, vol. 6518, pp. 89–99. Springer, Heidelberg (2011)
15. Woods, D., Chen, H.L., Goodfriend, S., Dabby, N., Winfree, E., Yin, P.: Active self-assembly of algorithmic shapes and patterns in polylogarithmic time. In: Proceedings of the 4th conference on Innovations in Theoretical Computer Science, ITCS 2013, pp. 353–354. ACM, New York (2013)

Probabilistic Reasoning with an Enzyme-Driven DNA Device

Iñaki Sainz de Murieta and Alfonso Rodríguez-Patón

Departamento de Inteligencia Artificial,
Universidad Politécnica de Madrid (UPM),
Campus de Montegancedo s/n, Boadilla del Monte 28660 Madrid, Spain
inaki.sainzdemurieta@upm.es, arpaton@fi.upm.es

Abstract. We present a biomolecular probabilistic model driven by the action of a DNA toolbox made of a set of DNA templates and enzymes that is able to perform Bayesian inference. The model will take single-stranded DNA as input data, representing the presence or absence of a specific molecular signal (the evidence). The program logic uses different DNA templates and their relative concentration ratios to encode the prior probability of a disease and the conditional probability of a signal given the disease. When the input and program molecules interact, an enzyme-driven cascade of reactions (DNA polymerase extension, nicking and degradation) is triggered, producing a different pair of single-stranded DNA species. Once the system reaches equilibrium, the ratio between the output species will represent the application of Bayes' law: the conditional probability of the disease given the signal. In other words, a qualitative diagnosis plus a quantitative degree of belief in that diagnosis. Thanks to the inherent amplification capability of this DNA toolbox, the resulting system will be able to to scale up (with longer cascades and thus more input signals) a Bayesian biosensor that we designed previously.

1   Introduction

Dynamic DNA nanotechnology is one of the areas of biomolecular computing that has developed most over the past decade. Many different models of DNA processors have been implemented since Adleman's seminal work [1]. We can find examples of DNA automata driven by restriction enzymes [2], deoxyribozyme-based DNA automata [3,4], DNA polymerase-based computers [5] or strand displacement circuits [6,7,8,9,10,11,12].

Most of the above models are designed as "use once" devices. This is a consequence of their operating principle: a set of molecules in a non-equilibrium state undertaking reactions and conformational changes until they reach a practically irreversible equilibrium state. Although this feature seems to be consistent with the objectives of structural DNA nanotechnology (e.g. DNA origami [13]), when we move to dynamic DNA nanotechnology the "use once" feature is a drawback rather than an advantage. Although they can still have very interesting applications (e.g. *in vitro* sensors and genetic diagnosis), every computation would

D. Soloveichik and B. Yurke (Eds.): DNA 2013, LNCS 8141, pp. 160–173, 2013.
© Springer International Publishing Switzerland 2013

require a new DNA device. In order to achieve more complex behaviors, such as bistability or oscillations, biomolecular computing models need to be driven by a continuous input flux of energy [14]. This could be achieved, for example, by the DNAzyme-driven [3,4] and catalytic enzyme-free [10,11] models cited above, as long as there is a continuous supply of input ribonucleated strands and fuel strands, respectively, in the environment (e.g. in an open reactor). The design of other biomolecular computing models depends fundamentally on the existence of an input energy flux. For example, RNA computers work with a continuous supply of NTP, used by RNA polymerase as fuel in the transcription process [15,16].

DNA polymerase was one of the first computational primitives used in the early models of DNA computing [1,17]. It was therefore not surprising to find it in the first autonomous DNA computer model: the Whiplash machine [5]. However, after that milestone, DNA polymerase-driven models remained outside the mainstream for years, mainly due to the need for thermal cycles. Interest in this topic rekindled after some breakthroughs exploiting isothermal DNA amplification protocols [18], such as an improved Whiplash model [19] or the DNA toolbox developed by Rondelez's team [20,21,22].

The DNA toolbox is specially interesting due to its similarities with RNA computers: it is also driven by a continuous supply of NTP, which is used to extend input DNA strands and produce output strands. It has recently led to impressive achievements, such as reliable oscillations [20], bistability [21] or population dynamics models like predator-prey [22]. Its operation is based on the action of a set of enzymes (DNA polymerase, an isothermal DNA nicking enzyme and a single-strand specific exonuclease) on the input strands and a set of single-stranded DNA templates, enabling the following set of basic reactions (see Figure 1):

- *Polymerization and nicking.* After the hybridization of an input DNA strand $\vec{A}$[1] at the 3' end of a DNA template $\overleftarrow{AB}$, DNA polymerase produces the double strand $\overleftrightarrow{AB}$. Since the duplex $\overleftrightarrow{A}$ contains the recognition sequence of the nicking enzyme, the newly polymerized strand is cleaved in two fragments $\vec{A}$ and $\vec{B}$, which will dissociate from the template due to their shorter length. $\vec{B}$ can also be displaced by further DNA polymerase activity. As result of this process, the input strands $\vec{A}$ periodically generate new strands $\vec{B}$ (see left panel in Figure 1).
- *Inactivation.* A special type of input DNA strand $\vec{B}$ can be used to inactivate a template $\overleftarrow{DE}$. $\vec{B}$ does not fully bind the recognition sequence of the nicking enzyme in the template, and since it is longer than the regular inputs $\vec{D}$, $\vec{B}$ wins the competition to bind the template almost irreversibly. Moreover, its 3' end does not bind the template, avoiding the action of DNA polymerase (see right bottom panel in Figure 1).

---

[1] A DNA strand denoted $\vec{A}$ is supposed to be Watson-Crick complementary to a DNA strand denoted $\overleftarrow{A}$, and would form a duplex $\overleftrightarrow{A}$ when both molecules hybridize.

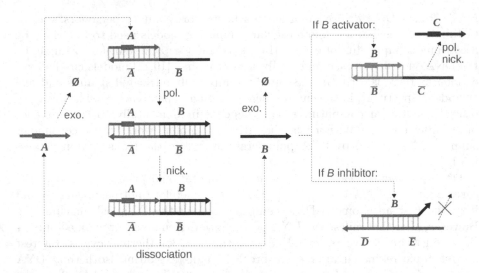

Fig. 1. DNA toolbox. The left panel shows the basic catalytic operation of the tool-box: the input strand $\overrightarrow{A}$ binds the 3' end of the DNA template $\overleftarrow{AB}$, allowing DNA polymerase to extend it forming the duplex $\overleftrightarrow{AB}$. Then the enzyme nickase binds to its recognition sequence in $\overleftrightarrow{A}$ (bold line) and cleaves the newly polymerized upper strand $\overrightarrow{AB}$ in two fragments $\overrightarrow{A}$ and $\overrightarrow{B}$, which can either dissociate from the template due to their shorter length, or let $\overrightarrow{B}$ be displaced by a new DNA polymerization of $\overrightarrow{A}$. The right panel shows the two possible operating modes of the output $\overrightarrow{B}$: as an activator it will enable the polymerization of another DNA strand $\overrightarrow{C}$ (see the motif at the top); as an inhibitor, it would bind in the middle of a DNA template $\overleftarrow{DE}$, inhibiting nicking and polymerization. All the DNA strands except the templates are subject to periodic degradation (see arrows pointing to ϕ).

– *Degradation.* Species dynamically generated by DNA polymerase are de-graded by a single-strand specific exonuclease. DNA templates are protected from the action of the exonuclease thanks to DNA backbone modifications at their 5' end.

Inspired by recent works presented above by Rondelez's team, we have identi-fied their DNA toolbox as an alternative to implementing probabilistic reasoning, which can be used when we want to consider diagnostic accuracy or uncertainty of tests in our clinical decisions (i.e., classic systems like Mycin [23]). With the aim of designing a model that can process this uncertainty, this article presents a Bayesian biosensor that reasons probabilistically and whose output represents the probability (value between 0 and 1) of a disease. Such a device can be used to estimate and update the probability of any diagnosis based in the light of new evidence, i.e., the presence or absence of a new specific signal (or set of signals). The DNA sensor device encodes two different probabilities as program data: the conditional probability of the signal given the disease ($P(signal|disease)$) and the prior probability of the disease ($P(disease)$). Then, when the sensor interacts with an input representing the evidence of a signal (its presence or absence), Bayes' law is autonomously computed by means of enzymatic reaction cascades, releasing a set of DNA species whose concentration ratio encodes the posterior probability of the disease given the input ($P(disease|signal)$). We presented a similar model in [24], which used DNA strand displacement instead of Rondelez's DNA toolbox.

The rest of the chapter is structured as follows. Section 2 includes an example of Bayesian inference that can be performed with the model. Sections 3, 4 and 5 show the encoding of input signals and prior and conditional probabilities, respectively. Section 6 details how the model implements the Bayesian inference process. Finally, Section 8 summarizes the conclusions and future work.

2   Example of Bayesian Inference

This section describes a basic Bayesian inference example.

Let us imagine that we want to diagnose whether a patient is affected by a certain disease d, whose possible diagnosis is "disease present" (D_1) or "disease absent" (D_0).

Based on empirical data, we can know upfront the prior probability of the disease. For this example, we consider both diagnoses to be equiprobable, which is represented as follows:

$$P(d) = \langle P(D = present), P(D = absent) \rangle = \langle P(D_1), P(D_0) \rangle = \langle 0.5, 0.5 \rangle.$$

Studying already diagnosed cases of this disease and its symptoms s (working as input signals), we can also ascertain upfront the conditional probability of a certain symptom (or signal) s given the disease d, $P(s|d)$:

$$P(S = absent)|D = absent) = P(S_0|D_0) = 0.7$$
$$P(S = present)|D = absent) = P(S_1|D_0) = 0.3$$
$$P(S = absent)|D = present) = P(S_0|D_1) = 0.2$$
$$P(S = present)|D = present) = P(S_1|D_1) = 0.8.$$

Now we test whether the patient has symptom s, which we interpret as a confirmation that the signal s is present (S_1). In the light of this new evidence, we can update our knowledge on the probability of the disease being present given that the signal is present, $P(D_1|S_1)$, applying the Bayes' law:

$$P(d|s) = \frac{P(s|d) \cdot P(d)}{P(s)} = \alpha \cdot P(s|d) \cdot P(d). \tag{1}$$

Since we do not know the prior probability of the signal $P(s)$, we can apply the second derivation of Bayes' law as stated in Equation 1:

$$P(D_1|S_1) = \alpha \cdot P(S_1|D_1) \cdot P(D_1) = \alpha \cdot 0.8 \cdot 0.5 = \alpha \cdot 0.4.$$
In order to find α, we need to calculate $P(D_0|S_1)$ as well:

$$P(D_0|S_1) = \alpha \cdot P(S_1|D_0) \cdot P(D_0) = \alpha \cdot 0.3 \cdot 0.5 = \alpha \cdot 0.15.$$

According to the foundations of probability theory, we know $P(D_1|S_1) + P(D_0|S_1) = 1$. We can use this knowledge to derive $\alpha = 1.81$ and $P(D_1|S_1) = 0.73$.

The biomolecular probabilistic inference devices described in the next sections of the paper can autonomously update their output probability values, such that they match the inference steps described in this example.

3   Encoding Input Evidences

Normally, a biomolecular device that senses real samples expecting a certain input signal In would reason as follows: if molecules In are present, the signal is present; otherwise the signal is absent. However, the devices that we propose use a different type of input logic, where the presence and absence of the signal are represented by the presence of different DNA species.

Thus, our input evidence is encoded using single-stranded DNA. A strand S_1 encodes the presence of an input signal, whereas a strand S_0 encodes the absence of the signal. As we are dealing with evidences, only one species can be present at a time: either S_1 (meaning the signal is present) or S_0 (meaning the signal is not present). These input signals will tell the sensor that the prior probability of the disease needs to be updated according to the given evidence.

However, if the system is to be able to deal with real biological samples, it needs to translate the presence of an external input signal In into strands S_1 (meaning input present in our system) and the absence of In into strands S_0 (meaning input absent in our system). A recent bistable implementation using this DNA toolbox illustrated an excellent way of translating the respective signals to produce strands S_1 and S_0 [21]. In this paper, a bistable switch producing a certain type of DNA species (which could be our S_0) in the absence of a certain type of input species In switched to producing another type of DNA species (which could be our S_1) in the presence of In. This model meets all the requirements to encode input evidence in the fashion described above. See [21] for details, which are omitted here due to space constraints.

4   Encoding Prior Probabilities

As illustrated by the example of Section 2, the prior probability of a disease is represented by the duple $P(d) = \langle P(D_1), P(D_0) \rangle$. Our model will use two different single-stranded DNA species to encode each possible probability value: $\overrightarrow{D_1}$ species representing $P(D = present)$ and $\overrightarrow{D_0}$ representing $P(D = absent)$. These strands will be produced from two DNA templates, $\overrightarrow{D_1 D_1}$ and $\overleftarrow{D_0 D_0}$. When $\overrightarrow{D_i}$ strands interact with their respective $\overleftarrow{D_i D_i}$ templates, $\overrightarrow{D_i}$ production increases (see Figure 2). At the same time, exonuclease degrades the production of $\overrightarrow{D_i}$ at a certain rate. The equations below govern this behavior:

$$\overrightarrow{D_i} + \overleftarrow{D_i D_i} \underset{k_a^{D_i} \cdot K_d^{D_i}}{\overset{k_a}{\rightleftharpoons}} \overrightarrow{D_i} \cdot \overleftarrow{D_i D_i} \xrightarrow{k_{cat}^{D_i}} \overrightarrow{D_i} + \overrightarrow{D_i} + \overleftarrow{D_i D_i} \qquad (2)$$

$$\overrightarrow{D_i} \xrightarrow{k_{dec}^{D_i}} \phi \,, \qquad (3)$$

where k_a is the association constant of $\overrightarrow{D_i}$, $K_d^{D_i}$ is the dissociation constant of $\overrightarrow{D_i}$ from the DNA template, $k_{dec}^{D_i}$ is the degradation rate of $\overrightarrow{D_i}$, and $k_{cat}^{D_i}$ is the rate of production of new strands $\overrightarrow{D_i}$. The constant really includes several reactions (polymerization, nicking and dissociation), but is confined to one here for reasons of space. Therefore, we would expect $k_{cat}^{D_i} \ll k_a^{D_i}$ and thus the respective Michaelis-Menten constant of the catalysis reaction would be $K_m^{D_i} \simeq K_d^{D_i}$ ($K_m^{D_i} = (k_a^{D_i} \cdot K_d^{D_i} + k_{cat}^{D_i})/k_a^{D_i}$).

When the system reaches equilibrium, the ratio between the concentration of both species will encode the prior probability, such that

$$P(D_i) = \frac{[\overrightarrow{D_i}]^{EQ}}{\sum\limits_{i=0}^{1} [\overrightarrow{D_i}]^{EQ}} = \frac{[\overrightarrow{D_i}]^{EQ}}{\lambda}, \tag{4}$$

where λ represents the sum of $[\overrightarrow{D_0}]$ and $[\overrightarrow{D_1}]$ that encodes the maximum probability 1. Each equilibrium concentration $[\overrightarrow{D_i}]^{EQ}$ is a function of the initial concentration of the templates $\overrightarrow{D_i D_i}$. Section 6 shows the derivation of this function.

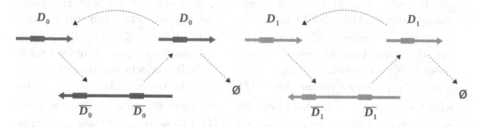

Fig. 2. Encoding prior probabilities. Thick regions of the strands represent the nickase recognition sequence. When a strand $\overrightarrow{D_i}$ at the top of the figure binds a template strand $\overleftarrow{D_i D_i}$ at the bottom of the figure, they form a complex $\overrightarrow{D_i}:\overleftarrow{D_i D_i}$ then DNA polymerase extends the upper strand to form the duplex $\overrightarrow{D_i D_i}:\overleftarrow{D_i D_i}$ and finally the enzyme nickase cleaves the newly polymerized strand in the middle. After the $\overrightarrow{D_i}$ strands dissociate from the template due to their short length, they can either be degraded by the exonuclease (arrows pointing to ϕ) or be recruited again by the template to produce more strands $\overrightarrow{D_i}$.

5   Encoding Conditional Probabilities

Conditional probabilities require the encoding of four different probability values: $P(S_0|D_0)$, $P(S_0|D_1)$, $P(S_1|D_0)$ and $P(S_1|D_1)$. Two different types of DNA templates will be used in the encoding of each probability value (see left side of Figure 3):

– Templates with format $\overleftarrow{D_i:D_i \wedge S_j}$ produce species $\overrightarrow{D_i \wedge S_j}$ in the presence of input strands $\overrightarrow{D_i}$ (see Figure 3), such that when the system reaches equilibrium $[\overrightarrow{D_i \wedge S_j}]^{EQ}$ is a function of $[\overrightarrow{D_i}]^{EQ}$ and $[\overleftarrow{D_i:D_i \wedge S_j}]$. The relative concentration of the templates with format $\overleftarrow{D_i : D_i \wedge S_j}$ encodes each conditional probability value, such that:

$$P(S_j|D_i) = \frac{\beta_{ij} \cdot [\overleftarrow{D_i : D_i \wedge S_j}]}{\sum\limits_{j=0}^{1} \beta_{ij} \cdot [\overleftarrow{D_i : D_i \wedge S_j}]} \tag{5}$$

$$\sum_{j=0}^{1} \beta_{ij} \cdot [\overleftarrow{D_i : D_i \wedge S_j}] = \sum_{j=0}^{1} \beta_{kj} \cdot [\overleftarrow{D_k : D_k \wedge S_j}] = \gamma \,, \; k \neq i \,, \tag{6}$$

where β_{ij} is a normalization coefficient and γ is the total normalized concentration of strands $[\overleftarrow{D_i : D_i \wedge S_j}]$ that represents probability 1. Section 6 will show the meaning of the β_{ij} coefficients and how $[\overrightarrow{D_i \wedge S_j}]^{EQ}$ is proportional to the product of $[\overrightarrow{D_i}]^{EQ}$ and $[\overleftarrow{D_i : D_i \wedge S_j}]$.

- Templates with format $\overrightarrow{D_i \wedge S_j : D_i'}$ have a twofold objective. First, they generate the output species D_i', whose relative concentration will encode the posterior probability of the disease given the signal ($P(d|s)$). Second, in conjunction with the input signal species S_i, they select what posterior probability computation should be produced as output: when the input signal is S_1 (S_0), it binds and inactivates the strands $\overleftarrow{D_i \wedge S_0 : D_i'}$ ($\overleftarrow{D_i \wedge S_1 : D_i'}$) (see the crossed-out arrows in Figure 3), so that there is only one source of species D_1' and another of D_0', whose ratio will conform the output probability: the posterior probability $P(D_i|S_j)$ of the disease. All the templates with format $\overleftarrow{D_i \wedge S_j : D_i'}$ must have the same concentration, so that there are no changes of relative proportions from $[\overrightarrow{D_i'}]$ in relation to their respective source $[\overrightarrow{D_i \wedge S_j}]$.

The equations below govern the behaviour of these components:

$$\overrightarrow{D_i} + \overleftarrow{D_i : D_i \wedge S_j} \underset{K_d^{D_i} \cdot k_a^{D_i}}{\overset{k_a}{\rightleftharpoons}} \overrightarrow{D_i} \cdot \overleftarrow{D_i : D_i \wedge S_j} \xrightarrow{k_{cat}^{D_i \wedge S_j}} \overrightarrow{D_i} + \overleftarrow{D_i : D_i S_j} + \overrightarrow{D_i \wedge S_j} \tag{7}$$

$$\overrightarrow{D_i \wedge S_j} + \overleftarrow{D_i \wedge S_j : D_i'} \underset{K_d^{D_i \wedge S_j} \cdot k_a^{D_i \wedge S_j}}{\overset{k_a}{\rightleftharpoons}} \overrightarrow{D_i \wedge S_j} \cdot \overleftarrow{D_i \wedge S_j : D_i'} \xrightarrow{k_{cat}^{D_i'}} \ldots$$

$$\ldots \xrightarrow{k_{cat}^{D_i'}} \overrightarrow{D_i \wedge S_j} + \overleftarrow{D_i \wedge S_j : D_i'} + \overrightarrow{D_i'} \tag{8}$$

$$\overrightarrow{S_k} + \overleftarrow{D_i \wedge S_j : D_i'} \underset{K_d^{S_k} \cdot k_a^{S_k}}{\overset{k_a}{\rightleftharpoons}} \overrightarrow{S_k} \cdot \overleftarrow{D_i \wedge S_j : D_i'} \,, \; k \neq j \tag{9}$$

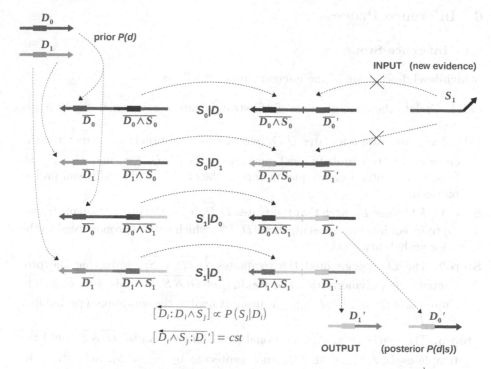

Fig. 3. Encoding conditional probabilities. The prior probability strands $\overrightarrow{D_i}$ bind the templates on the left side, enabling the production of $\overleftarrow{D_i \wedge S_j}$ strands (via polymerization and nicking) used to encode conditional probability. These strands will then activate the templates on the right side not protected by the input strands S_j (see the crossed-out arrows), producing the output strands $\overrightarrow{D_i}'$, whose concentration ratio encodes the posterior probability $P(d|s)$.

$$\overrightarrow{D_i \wedge S_j} \xrightarrow{k_{dec}^{D_i \wedge S_j}} \phi \tag{10}$$

$$\overrightarrow{D_i}' \xrightarrow{k_{dec}^{D_i'}} \phi \tag{11}$$

$$\overrightarrow{S_j} \xrightarrow{k_{dec}^{S_j}} \phi , \tag{12}$$

where k_a is the association rate of $\overrightarrow{D_i}$, $\overrightarrow{D_i \wedge S_j}$ and $\overrightarrow{S_k}$; $K_d^{D_i}$, $K_d^{D_i \wedge S_j}$ and $K_d^{S_k}$ are their respective dissociation constants; $k_{cat}^{D_i S_j}$ and $k_{cat}^{D_i'}$ are the production rates of strands $\overrightarrow{D_i \wedge S_j}$ and $\overrightarrow{D_i}'$; $k_{dec}^{D_i \wedge S_j}$, $k_{dec}^{D_i'}$ and $k_{dec}^{S_j}$ are the degradation constants of $\overrightarrow{D_i \wedge S_j}$ and $\overrightarrow{D_i}'$ and S_j.

6   Inference Process

6.1   Inference Steps

A high-level description of the inference process follows:

Goal. Update the concentration of $\overrightarrow{D_i}'$ strands once a new signal ($\overrightarrow{S_0}$ or $\overrightarrow{S_1}$) is detected.

Initial set-up. Add templates $\overleftarrow{D_i D_i}$ (whose concentration is a parameter in the encoding of prior probabilities), and templates $\overleftarrow{D_i : D_i \wedge S_j}$ and $\overleftarrow{D_i \wedge S_j : D_i'}$ (whose concentrations are parameters in the encoding of conditional probabilities).

Step 1. Add some $\overrightarrow{D_i}$, such that templates $\overleftarrow{D_i D_i}$ bring the production of strands $\overrightarrow{D_i}$ to its equilibrium concentration $[\overrightarrow{D_i}]^{EQ}$, which will be proportional to the prior probability $P(D_i)$.

Step 2. The $\overrightarrow{D_i}$ species bind the templates $\overleftarrow{D_i : D_i \wedge S_j}$, activating the production (via polymerization and nicking) of $\overrightarrow{D_i \wedge S_j}$ strands, whose equilibrium concentration $[\overrightarrow{D_i \wedge S_j}]^{EQ}$ is proportional to the conditional probability $P(S_j|D_i)$.

Step 3. The newly created "conditional probability strands" $\overrightarrow{D_i \wedge S_j}$ bind the templates $\overleftarrow{D_i \wedge S_j : D_i'}$ that are not protected by $\overrightarrow{S_0}$ or $\overrightarrow{S_1}$, activating the production (via polymerization and nicking) of the output species D_i'.

Read-out. The new concentration ratio of $\overrightarrow{D_i}'$ encodes the posterior probability $P(D_i|S_j)$.

This description is refined below providing a more thorough analysis of the process with estimations and derivations.

6.2   Modeling the Inference

From the equations presented in Sections 4 and 5, we can build a derivation that relates the output concentrations $[\overrightarrow{D_i}']$ to the initial concentrations of the strands encoding prior and conditional probabilities.

Based on Equations 2, 3 and the Michaelis-Menten model [25], we can infer how $[\overrightarrow{D_i}]$ changes in time (see Equation 13) and, applying the equilibrium condition $(d[\overrightarrow{D_i}]/dt = 0)$, obtain derivations for $[\overrightarrow{D_i}]^{EQ}$ (see Equation 14) and the initial $[\overleftarrow{D_i D_i}]$ (see Equation 15):

$$\frac{d[\overrightarrow{D_i}]}{dt} = \frac{k_{cat}^{D_i} \cdot [\overrightarrow{D_i}] \cdot [\overleftarrow{D_i D_i}]}{K_d^{D_i} + [\overrightarrow{D_i}]} - k_{dec}^{D_i} \cdot [\overrightarrow{D_i}] \tag{13}$$

$$[\overrightarrow{D_i}]^{EQ} = \frac{k_{cat}^{D_i}}{k_{dec}^{D_i}}[\overleftarrow{D_i D_i}] - K_d^{D_i} \tag{14}$$

$$[\overleftrightarrow{D_i D_i}] = \frac{k_{dec}^{D_i}}{k_{cat}^{D_i}}([\overrightarrow{D_i}]^{EQ} + K_d^{D_i}) \,. \tag{15}$$

A similar procedure can be applied for $\overrightarrow{D_i \wedge S_j}$ from Equations 7 and 10, obtaining a derivation for $[\overrightarrow{D_i \wedge S_j}]^{EQ}$ (see Equations 16 and 17). We are assuming $K_d^{D_i} >> [\overrightarrow{D_i}]$, which could be achieved with an appropriate temperature increase:

$$\frac{d[\overrightarrow{D_i \wedge S_j}]}{dt} = \frac{k_{cat}^{D_i \wedge S_j}}{K_d^{D_i}}[\overrightarrow{D_i}] \cdot [\overleftrightarrow{D_i : D_i S_j}] - k_{dec}^{D_i \wedge S_j} \cdot [\overrightarrow{D_i \wedge S_j}] \tag{16}$$

$$[\overrightarrow{D_i \wedge S_j}]^{EQ} = \frac{k_{cat}^{D_i \wedge S_j}}{K_d^{D_i} \cdot k_{dec}^{D_i \wedge S_j}}[\overrightarrow{D_i}] \cdot [\overleftrightarrow{D_i : D_i S_j}] \,. \tag{17}$$

The formulation of $[\overrightarrow{D_i'}]$ is a bit more intricate, because the Michaelis-Menten derivation needs to consider the interaction of the inhibiting input species S_i, which represses the catalysis. Based on Equations 8, 9 and 11, and also assuming $K_d^{D_i \wedge S_j} >> [\overrightarrow{D_i \wedge S_j}]$, we can infer $[\overrightarrow{D_i'}]^{EQ}$ (see Equations 18 and 19):

$$\frac{d[\overrightarrow{D_i'}]}{dt} = \sum_{j=0}^{1} \frac{k_{cat}^{D_i'} \cdot [\overrightarrow{D_i \wedge S_j}] \cdot [\overleftrightarrow{D_i \wedge S_j : D_i'}]}{K_d^{D_i \wedge S_j} \cdot (1 + \frac{[S_k^{k \neq j}]}{K_d^{S_k}})} - k_{dec}^{D_i'} \cdot [\overrightarrow{D_i'}] \tag{18}$$

$$[\overrightarrow{D_i'}]^{EQ} = \sum_{j=0}^{1} \frac{k_{cat}^{D_i'}}{K_d^{D_i \wedge S_j} \cdot k_{dec}^{D_i'} \cdot (1 + \frac{[S_k^{k \neq j}]}{K_d^{S_k}})}[\overrightarrow{D_i \wedge S_j}] \cdot [\overleftrightarrow{D_i \wedge S_j : D_i'}] \,. \tag{19}$$

Taking into account that species S_0 and S_1 are never present at the same time, $[\overrightarrow{S_i}] >> [\overrightarrow{D_i \wedge S_j}] + [\overleftrightarrow{D_i \wedge S_j : D_i'}]$ and $K_d^{S_k} << K_d^{D_i \wedge S_j}$, we can neglect the terms of the sum in Equation 19 where $[\overrightarrow{S_k}] > 0$ and derive a simpler expression for $[\overrightarrow{D_i'}]$:

$$[\overrightarrow{D_i'}]^{EQ}_{[S_j]=0,\, [S_k^{k \neq j}]>0} = \frac{k_{cat}^{D_i'}}{K_d^{D_i \wedge S_j} \cdot k_{dec}^{D_i'}}[\overrightarrow{D_i \wedge S_j}] \cdot [\overleftrightarrow{D_i \wedge S_j : D_i'}] \,. \tag{20}$$

Substituting Equation 17 in Equation 20, and reordering constant values to the left and variables to the right:

$$[\overrightarrow{D_i'}]^{EQ}_{[S_j]=0,\, [S_k^{k \neq j}]>0} = \frac{k_{cat}^{D_i'} \cdot k_{cat}^{D_i \wedge S_j} \cdot [\overleftrightarrow{D_i \wedge S_j : D_i'}]}{K_d^{D_i \wedge S_j} \cdot k_{dec}^{D_i'} \cdot K_d^{D_i} \cdot k_{dec}^{D_i \wedge S_j}}[\overleftrightarrow{D_i : D_i S_j}] \cdot [\overrightarrow{D_i}] =$$

$$= \beta_{ij} \cdot [\overleftrightarrow{D_i : D_i S_j}] \cdot [\overrightarrow{D_i}] \,. \tag{21}$$

In the above Equation 21, all the constant terms have been grouped in the parameter β_{ij} (already introduced in Equations 5 and 6). The term $[\overrightarrow{D_i}]$ is proportional to the prior probability (see Equation 4), and the product $\beta_{ij} \cdot [\overleftarrow{D_i : D_i S_j}]$ is proportional to the conditional probability (see Equation 5). The derivation below shows how the ratio between $[\overrightarrow{D_0'}]^{EQ}$ and $[\overrightarrow{D_1'}]^{EQ}$ determines posterior probability $P(d|s)$:

$$
\frac{[\overrightarrow{D_i'}]^{EQ}}{[\overrightarrow{D_0'}]^{EQ} + [\overrightarrow{D_1'}]^{EQ}} = \frac{\beta_{ij} \cdot [\overleftarrow{D_i : D_i S_j}] \cdot [\overrightarrow{D_i}]}{\beta_{0j} \cdot [\overleftarrow{D_0 : D_0 S_j}] \cdot [\overrightarrow{D_0}] + \beta_{1j} \cdot [\overleftarrow{D_1 : D_1 S_j}] \cdot [\overrightarrow{D_1}]} =
$$

$$
= \frac{\beta_{ij} \cdot \frac{\gamma}{\beta_{ij}} \cdot P(S_j|D_i) \cdot \lambda \cdot P(D_i)}{\beta_{0j} \cdot \frac{\gamma}{\beta_{0j}} \cdot P(S_j|D_0) \cdot \lambda \cdot P(D_0) + \beta_{1j} \cdot \frac{\gamma}{\beta_{1j}} \cdot P(S_j|D_1) \cdot \lambda \cdot P(D_1)} =
$$

$$
= \frac{P(S_j|D_i) \cdot P(D_i)}{P(S_j|D_0) \cdot P(D_0) + P(S_j|D_1) \cdot P(D_1)} = \frac{P(S_j|D_i) \cdot P(D_i)}{P(S_j)} = P(D_i|S_j) .
$$

7   Discussion

The DNA biosensor presented here operates as a Bayesian inference device. It is capable of introducing quantitative information, highlighted by the molecular indicators or signals, into the tests. It builds on our previous work [24], but uses the DNA toolbox recently introduced by Rondelez [20,21] instead of the DNA strand displacement operation. Another aim was to map the basic concepts of probability theory and Bayesian inference into the toolbox motifs, for use as design patterns when implementing Bayesian reasoning with DNA.

The example detailed in Section 6 has used only one input signal. For this model to have realistic applications in genetic diagnosis, however, it needs to deal with more than one signal ($s^1, ..., s^n$) for the same disease d (superscripts denote the signal number). According to Equation 1, the following formulation of Bayes' law would need to be solved: $P(d|s^1, ..., s^n) = \alpha \cdot P(d) \cdot P(s^1, ..., s^n|d)$. Assuming conditional independence of the signals given the disease (as in the naïve Bayes model [26]) we can derive the following expression: $P(d|s^1, ..., s^n) = \alpha \cdot P(d) \cdot P(s^1|d) \cdot ... \cdot P(s^n|d)$, meaning the initial probability statement with multiple input signals can be decomposed into conditional probability products, which can be encoded by cascading the devices presented here.

This research has addressed the two main improvement opportunities of the work that we presented elsewhere [24]:

Reusability. Devices are conceived for just one use. If the inputs are altered after the output signals become stable, the new output would not be correct any more. We would need a new initialised set of devices to deal with a new input. This research solves this problem with the action of the single-strand specific exonuclease, which periodically degrades all the non-template strands not protected at their 5' end. This way, when the initial intput data flux S_k is stopped in favor of the new flux $S_{k'}$ ($k \neq k'; k, k' \in 0..1$), the system will converge to the total elimination of S_k (since that species is only degraded and not replenished). The same should happen with the intermediate species $\overleftarrow{D_i \wedge S_j}$ and output species $\overleftarrow{D_i'}$, driving the system to converge to a the correct output for input $S_{k'}$.

Signal attenuation. In theory, the model in [24] was also able to deal with multiple input signals by cascading the outputs as inputs of other conditional probability devices downstream. However, each inference iteration would attenuate the signal by an average of 50%. The replacement of strand displacement by an enzymatic catalysis, with inherent amplification capabilities, overcomes this drawback allowing longer inference cascades and thus more input signals.

8   Conclusions and Future Work

We have designed a biomolecular probabilistic expert system for genetic diagnosis. This is an enzyme-driven DNA device able to:

1. Encode diagnostic probabilistic information in single-stranded DNA.
2. Sense DNA inputs.
3. Process probabilistic information, encoded either as a steady state concentration of single-stranded DNA (for prior probabilities) or as a fixed concentration of single-stranded DNA (for conditional probabilities).
4. Release output molecules (duples of single-stranded DNA encoding a probability proportional to their concentration ratio).
5. Update the probability of the disease depending on the different single-stranded DNA inputs detected following Bayes' rule.

The model is autonomous and can be implemented according to the DNA toolbox presented in [20,21]. We think this and the other model that we introduced in [24] have the potential to deliver new quantitative applications of probabilistic genetic diagnosis *in vitro*. We plan to build, improve and generalize both models in a wet lab to work with all types of Bayesian networks (and not just naïve Bayes approaches [26]).

Acknowledgments. This research was partially supported by project BAC-TOCOM funded by a European Commission 7th Framework Programme grant (FET Proactive area) and by Spanish Ministry of Finance project TIN2012-36992.

References

1. Adleman, L.M.: Molecular computation of solutions to combinatorial problems. Science 266(5187), 1021–1024 (1994)
2. Benenson, Y., Gil, B., Ben-Dor, U., Adar, R., Shapiro, E.: An autonomous molecular computer for logical control of gene expression. Nature 429(6990), 423–429 (2004)
3. Stojanovic, M.N., Stefanovic, D.: A deoxyribozyme-based molecular automaton. Nature Biotechnology 21(9), 1069–1074 (2003)
4. Pei, R., Matamoros, E., Liu, M., Stefanovic, D., Stojanovic, M.N.: Training a molecular automaton to play a game. Nature Nanotechnology 5(11), 773–777 (2010)
5. Hagiya, M., Arita, M., Kiga, D., Sakamoto, K., Yokoyama, S.: Towards Parallel Evaluation and Learning of Boolean μ-Formulas with Molecules 48, 105–114 (1997)
6. Yurke, B., Turberfield, A.J., Mills, A.P., Simmel, F.C., Neumann, J.L.: A DNA-fuelled molecular machine made of DNA. Nature 406(6796), 605–608 (2000)
7. Seelig, G., Soloveichik, D., Zhang, D.Y., Winfree, E.: Enzyme-Free Nucleic Acid Logic Circuits. Science 314(5805), 1585–1588 (2006)
8. Rodríguez-Patón, A., de Murieta, I.S., Sosík, P.: Autonomous resolution based on DNA strand displacement. In: Cardelli, L., Shih, W. (eds.) DNA 17. LNCS, vol. 6937, pp. 190–203. Springer, Heidelberg (2011)
9. Sainz de Murieta, I., Rodríguez-Patón, A.: DNA biosensors that reason. Biosystems 109(2), 91–104 (2012)
10. Qian, L., Winfree, E.: Scaling up digital circuit computation with DNA strand displacement cascades. Science 332(6034), 1196–1201 (2011)
11. Qian, L., Winfree, E., Bruck, J.: Neural network computation with DNA strand displacement cascades. Nature 475(7356), 368–372 (2011)
12. Soloveichik, D., Seelig, G., Winfree, E.: DNA as a universal substrate for chemical kinetics. Proceedings of the National Academy of Sciences 107(12), 5393–5398 (2010)
13. Rothemund, P.W.K.: Folding DNA to create nanoscale shapes and patterns. Nature 440(7082), 297–302 (2006)
14. Kjelstrup, S., Bedeaux, D.: Non-Equilibrium Thermodynamics of Heterogeneous Systems. Series on Advances in Statistical Mechanics. World Scientific (2008)
15. Benenson, Y.: Synthetic biology with RNA: progress report. Current Opinion in Chemical Biology 16(3-4), 278–284 (2012)
16. Weitz, M., Simmel, F.C.: Synthetic in vitro transcription circuits. Transcription 3(2), 87–91 (2012)
17. Amos, M.: Theoretical and Experimental DNA Computation. Natural computing series. Springer, Heidelberg (2005)
18. Walker, G.T., Little, M.C., Nadeau, J.G., Shank, D.D.: Isothermal in vitro amplification of DNA by a restriction enzyme/DNA polymerase system. Proceedings of the National Academy of Sciences 89(1), 392–396 (1992)
19. Reif, J., Majumder, U.: Isothermal reactivating whiplash PCR for locally programmable molecular computation. Natural Computing 9, 183–206 (2010)
20. Montagne, K., Plasson, R., Sakai, Y., Fujii, T., Rondelez, Y.: Programming an in vitro DNA oscillator using a molecular networking strategy. Molecular Systems Biology 7(1) (2011)

21. Padirac, A., Fujii, T., Rondelez, Y.: Bottom-up construction of in vitro switchable memories. Proceedings of the National Academy of Sciences 109(47), E3212–E3220 (2012)
22. Fujii, T., Rondelez, Y.: Predator-prey molecular ecosystems. ACS Nano 7(1), 27–34 (2013)
23. Shortliffe, E.H., Buchanan, B.G.: A model of inexact reasoning in medicine. Mathematical Biosciences 23(3-4), 351–379 (1975)
24. Sainz de Murieta, I., Rodríguez-Patón, A.: Probabilistic reasoning with a bayesian DNA device based on strand displacement. In: Stefanovic, D., Turberfield, A. (eds.) DNA 2012. LNCS, vol. 7433, pp. 110–122. Springer, Heidelberg (2012)
25. Johnson, K.A., Goody, R.S.: The original michaelis constant: Translation of the, michaelis-menten paper. Biochemistry 50(39), 8264–8269 (1913)
26. Minsky, M.: Steps toward artificial intelligence. Proceedings of the IRE 49(1), 8–30 (1961)

Staged Self-assembly and Polyomino Context-Free Grammars*

Andrew Winslow**

Department of Computer Science, Tufts University
awinslow@cs.tufts.edu

Abstract. Previous work by Demaine et al. (2012) developed a strong connection between smallest context-free grammars and staged self-assembly systems for one-dimensional strings and assemblies. We extend this work to two-dimensional polyominoes and assemblies, comparing staged self-assembly systems to a natural generalization of context-free grammars we call *polyomino context-free grammars (PCFGs)*.

We achieve nearly optimal bounds on the largest ratios of the smallest PCFG and staged self-assembly system for a given polyomino with n cells. For the ratio of PCFGs *over* assembly systems, we show that the smallest PCFG can be an $\Omega(n/\log^3 n)$-factor larger than the smallest staged assembly system, even when restricted to square polyominoes. For the ratio of assembly systems over PCFGs, we show that the smallest staged assembly system is never more than a $O(\log n)$-factor larger than the smallest PCFG and is sometimes an $\Omega(\log n/\log \log n)$-factor larger.

1 Introduction

In the mid-1990s, the Ph.D. thesis of Erik Winfree [14] introduced a theoretical model of self-assembling nanoparticles. In this model, which he called the *abstract tile assembly model (aTAM)*, square particles called *tiles* attach edgewise to each other if their edges share a common *glue* and the bond strength is sufficient to overcome the kinetic energy or *temperature* of the system. The products of these systems are *assemblies*: aggregates of tiles forming via crystal-like growth starting at a *seed tile*. Surprisingly, these tile systems have been shown to be computationally universal [14,5], self-simulating [8,9], and capable of optimally encoding arbitrary shapes [12,1,13].

In parallel with work on the aTAM, a number of variations on the model have been proposed and investigated. One well-studied variant called the *hierarchical* [4] or *two-handed assembly model (2HAM)* [6] eliminates the seed tile and allows tiles and assemblies to attach in arbitrary order. This model was shown to be capable of (theoretically) faster assembly of squares [4] and simulation of aTAM systems [2], including capturing the seed-originated growth dynamics. A generalization of the 2HAM model proposed by Demaine et al. [6] is the *staged*

---

* A full version of this paper can be found at http://arxiv.org/abs/1304.7038
** Supported in part by National Science Foundation grant CBET-0941538.

D. Soloveichik and B. Yurke (Eds.): DNA 2013, LNCS 8141, pp. 174–188, 2013.
© Springer International Publishing Switzerland 2013

assembly model, which allows the assemblies produced by one system to be used as reagents (in place of tiles) for another system, yielding systems divided into sequential assembly *stages*. They showed that such sequential assembly systems can replace the role of glues in encoding complex assemblies, allowing the construction of arbitrary shapes efficiently while only using a constant number of glue types, a result impossible in the aTAM or 2HAM.

To understand the power of the staged assembly model, Demaine et al. [7] studied the problem of finding the smallest system producing a one-dimensional assembly with a given sequence of labels on its tiles, called a *label string*. They proved that for systems with a constant number of glue types, this problem is equivalent to the well-studied problem of finding the smallest context-free grammar whose language is the given label string, also called the *smallest grammar problem* (see [11,3]). For systems with unlimited glue types, they proved that the ratio of the smallest context-free grammar *over* the smallest system producing an assembly with a given label string of length n (which they call *separation*) is $\Omega(\sqrt{n/\log n})$ and $O((n/\log n)^{2/3})$ in the worst case.

In this paper we consider the two-dimensional version of this problem: finding the smallest staged assembly system producing an assembly with a given *label polyomino*. For systems with constant glue types and no cooperative bonding, we achieve separation of grammars *over* these systems of $\Omega(n/(\log \log n)^2)$ for polyominoes with n cells (Sect. 6.1), and $\Omega(n/\log^3 n)$ when restricted to rectangular (Sect. 6.2) or square (Sect. 6.3) polyominoes with a constant number of labels. Adding the restriction that each step of the assembly process produces a single product, we achieve $\Omega(n/\log^3 n)$ separation for general polyominoes with a single label (Sect. 6.1). For the separation of staged assembly systems *over* grammars, we achieve bounds of $\Omega(\log n/\log \log n)$ (Sect. 4) and, constructively, $O(\log n)$ (Sect. 5). For all of these results, we use a simple definition of context-free grammars on polyominoes that generalizes the deterministic context-free grammars (called *RCFGs*) of [7].

When taken together, these results give a nearly complete picture of how smallest context-free grammars and staged assembly systems compare. For some polyominoes, staged assembly systems are exponentially smaller than context-free grammars ($O(\log n)$ vs. $\Omega(n/\log^3 n)$). On the other hand, given a polyomino and grammar deriving it, one can construct a staged assembly system that is a (nearly optimal) $O(\log n)$-factor larger and produces an assembly with a label polyomino replicating the polyomino.

2   Staged Self-assembly

An instance of the staged tile assembly model is called a *staged assembly system* or *system*, abbreviated *SAS*. A SAS $S = (T, G, \tau, M, B)$ is specified by five parts: a *tile set* T of square *tiles*, a *glue function* $G : \Sigma(G)^2 \to \{0, 1, \ldots, \tau\}$, a *temperature* $\tau \in \mathbb{N}$, a directed acyclic *mix graph* $M = (V, E)$, and a *start bin function* $B : V_L \to T$ from the *leaf vertices* $V_L \subseteq V$ of M with no incoming edges.

Each tile $t \in T$ is specified by a 5-tuple (l, g_n, g_e, g_s, g_w) consisting of a label l taken from an alphabet $\Sigma(T)$ (denoted $l(t)$) and a set of four non-negative integers in $\Sigma(G) = \{0, 1, \ldots, k\}$ specifying the *glues* on the sides of t with normal vectors $\langle 0, 1 \rangle$ (north), $\langle 1, 0 \rangle$ (east), $\langle 0, -1, \rangle$ (south), and $\langle -1, 0 \rangle$ (west), respectively, denoted $g_u(t)$. In this work we only consider glue functions with the constraints that if $G(g_i, g_j) > 0$ then $g_i = g_j$, and $G(0, 0) = 0$.

A *configuration* is a partial function $C : \mathbb{Z}^2 \rightarrow T$ mapping locations on the integer lattice to tiles. Any two locations $p_1 = (x_1, y_1)$, $p_2 = (x_2, y_2)$ in the domain of C (denoted $\text{dom}(C)$) are *adjacent* if $\|p_2 - p_1\| = 1$ and the *bond strength* between any pair of tiles $C(p_1)$ and $C(p_2)$ at adjacent locations is $G(g_{p_2 - p_1}(C(p_1)), g_{p_1 - p_2}(C(p_2)))$. A *configuration* is a τ-*stable assembly* or an *assembly at temperature* τ if $\text{dom}(C)$ is connected on the lattice and, for any partition of $\text{dom}(C)$ into two subconfigurations C_1, C_2, the sum of the bond strengths between tiles at pairs of locations $p_1 \in \text{dom}(C_1)$, $p_2 \in \text{dom}(C_2)$ is at least τ. Any pair of configurations C_1, C_2 are *equivalent* if there exists a vector $v = \langle x, y \rangle$ such that $\text{dom}(C_1) = \{p + v \mid p \in \text{dom}(C_2)\}$ and $C_1(p) = C_2(p + v)$ for all $p \in \text{dom}(C_1)$. Two τ-stable assemblies A_1, A_2 are said to *assemble* into a *superassembly* A_3 if there exists a translation vector $v = \langle x, y \rangle$ such that $\text{dom}(A_1) \cap \{p + v \mid p \in A_2\} = \emptyset$ and A_3 defined by the partial functions A_1 and A_2' with $A_2'(p) = A_2(p + v)$ is a τ-stable assembly.

Each vertex of the mix graph M describes a *two-handed assembly process*. This process starts with a set of τ-stable *input assemblies* I. The set of *assembled assemblies* Q is defined recursively as $I \subseteq Q$, and for any pair of assemblies $A_1, A_2 \in Q$ with superassembly A_3, $A_3 \in Q$. Finally, the set of *products* $P \subseteq Q$ is the set of assemblies A such that for any assembly A', no superassembly of A and A' is in Q.

The mix graph $M = (V, E)$ of $\mathcal{S}$ defines a set of two-handed assembly processes (called *mixings*) for the non-leaf vertices of M (called *bins*). The input assemblies of the mixing at vertex v is the union of all products of mixings at vertices v' with $(v', v) \in E$. The start bin function B defines the lone single-tile product of each mixings at a leaf bin. The system $\mathcal{S}$ is said to *produce* an assembly A if some mixing of $\mathcal{S}$ has a single product, A. We define the size of $\mathcal{S}$, denoted $\mathcal{S}$, to be $|E|$, the number of edges in M. If every mixing in a $\mathcal{S}$ has a single product, then $\mathcal{S}$ is a *singular self-assembly system (SSAS)*.

The results of Section 6.4 use the notion of a self-assembly system $\mathcal{S}'$ *simulating* a system $\mathcal{S}$ by carrying out the same sequence of mixings and producing a set of scaled assemblies. Formally, we say a system $\mathcal{S}' = (T', G', \tau, M', B')$ *simulates* a system $\mathcal{S} = (T, G, \tau, M, B)$ at *scale* b if there exist two functions f, g with the following properties:

(1) The function $f : (\Sigma(T') \cup \{\varnothing\})^{b^2} \rightarrow \Sigma(T) \cup \{\varnothing\}$ maps the labels of $b \times b$ regions of tiles (called *blocks*) to a label of a tile in T. The empty label $\varnothing$ denotes no tile.
(2) The function $g : S' \rightarrow V$ maps a subset S' of the vertices of the mix graph M' to vertices of the mix graph M such that g is an isomorphism between the subgraph induced by S' in M' and the graph M.

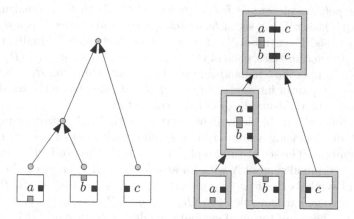

Fig. 1. A self-assembly system (SAS) consisting of a mix graph and tile types (left), and the assemblies produced by carrying out the algorithmic process of staged self-assembly (right).

(3) Let $P(v)$ be the set of products of the bin corresponding to vertex v in a mix graph. Then for each vertex $v \in M$ with $v' = g^{-1}(v)$, $P(v) = \{f(p) \mid p \in P(v')\}$.

Intuitively, f defines a correspondence between the b-scaled macrotiles in $\mathcal{S}'$ simulating tiles in $\mathcal{S}$, and g defines a correspondence between bins in the systems. Property (3) requires that f and g do, in fact, define correspondence between what the systems produce.

The self-assembly systems constructed in Sections 5 and 6 produce only *mismatch-free assemblies*: assemblies in which every pair of incident sides of two tiles in the assembly have the same glue. A system is defined to be *mismatch-free* if every product of the system is mismatch-free.

3 Polyomino Context-Free Grammars

Here we describe polyominoes, a generalization of strings, and polyomino context-free grammars, a generalization of deterministic context-free grammars. These objects replace the strings and restricted context-free grammars (RCFGs) of Demaine et al. [7].

A *labeled polyomino* or *polyomino* $P = (S, L)$ is defined by a connected set of points S on the square lattice (called *cells*) containing $(0, 0)$ and a label function $L : S \to \Sigma(P)$ mapping each cell of P to a *label* contained in an alphabet $\Sigma(P)$. The *size* of P is the number of cells P contains and is denoted $|P|$. The label of the cell at lattice point (x, y) is denoted $L((x, y))$ and we define $P(x, y) = L((x, y))$ for notational convenience. We refer to the *label* or *color* of a cell interchangeably.

Define a *polyomino context-free grammar (PCFG)* to be a quadruple $G = (\Sigma, \Gamma, S, \Delta)$. The set Σ is a set of *terminal symbols* and the set Γ is a set of *non-terminal symbols*. The symbol $S \in \Gamma$ is a special *start symbol*. Finally, the set Δ consists of *production rules*, each of the form $N \to (R_1, (x_1, y_1)) \ldots (R_j, (x_j, y_j))$ where $N \in \Gamma$ and is the left-hand side symbol of only this rule, $R_i \in N \cup T$, and each (x_i, y_i) is a pair of integers. The *size* of G is defined to be the total number of symbols on the right-hand sides of the rules of Δ.

A polyomino P can be derived by starting with S, the start symbol of G, and repeatedly replacing a non-terminal symbol with a set of non-terminal and terminal symbols. The set of valid replacements is Δ, the production rules of G, where a non-terminal symbol N with lower-leftmost cell at (x, y) can be replaced with a set of symbols R_1 at $(x+x_1, y+y_1)$, R_2 at $(x+x_2, y+y_2)$, $\ldots$, R_j at $(x+x_j, y+y_j)$ if there exists a rule $N \to (R_1, (x_1, y_1))(R_2, (x_2, y_2)) \ldots (R_j, (x_j, y_j))$. Additionally, the set of terminal symbol cells derivable starting with S must be connected and pairwise disjoint.

The polyomino P derived by the start symbol of a grammar G is called the *language of G*, denoted $L(G)$, and G is said to *derive* P. In the remainder of the paper we assume that each production rule has at most two right-hand side symbols (equivalent to binary normal form for 1D CFGs), as any PCFG can be converted to this form with only a factor-2 increase in size. Such a conversion is done by iteratively replacing two right-hand side symbols R_i, $R_{i'}$ with a new non-terminal symbol Q, and adding a new rule replacing Q with R_i and $R_{i'}$.

Intuitively, a polyomino context-free grammar is a recursive decomposition of a polyomino into smaller polyominoes. Because each non-terminal symbol is the left-hand side symbol of at most one rule, each non-terminal corresponds to a subpolyomino of the derived polyomino. Then each production rule is a decomposition of a subpolyomino into smaller subpolyominoes (see Figure 2).

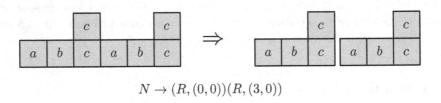

$$N \to (R, (0,0))(R, (3,0))$$

Fig. 2. Each production rule in a PCFG generating a single shape is a decomposition of the left-hand side non-terminal symbol's polyomino into the right-hand side symbols' polyominoes

In this interpretation, the smallest grammar deriving a given polyomino is equivalent to a decomposition using the fewest *distinct* subpolyominoes in the decomposition. As for the smallest CFG for a given string, the smallest PCFG for a given polyomino is deterministic and finding such a grammar is NP-hard. Moreover, even approximating the smallest grammar is NP-hard [3], and achieving optimal approximation algorithms remains open [10].

In Section 5 we construct self-assembly systems that produce assemblies whose label polyominoes are scaled versions of other polyominoes, with some amount of "fuzz" in each scaled cell. A polyomino $P' = (S', L')$ is said to be a (c, d)-*fuzzy replica* of a polyomino $P = (S, L)$ if there exists a vector $\langle x_t, y_t \rangle$ with the following properties:

1. For each block of cells $S'_{(i,j)} = \{(x, y) \mid x_t + di \le x < x_t + d(i+1), y_t + dj \le y < y_t + d(j+1)\}$ (called a *supercell*), $S'_{(i,j)} \cap S' \ne \varnothing$ if and only if $(i, j) \subseteq S$.
2. For each supercell $S'_{(i,j)}$ containing a cell of P', the subset of *label cells* $\{(x, y) \mid x_t + di + (d-c)/2 \le x < x_t + d(i+1) + (d-c)/2, y_t + dj + (d-c)/2 \le y < y_t + d(j+1) + (d-c)/2\}$ consists of c^2 cells of P', with all cells having identical label, called the *label of the supercell* and denoted $\mathcal{L}_{(i,j)}$.
3. For each supercell $S'_{(i,j)}$, any cell that is not a label cell of $S'_{(i,j)}$ has a common *fuzz label* in L'.
4. For each supercell $S'_{(i,j)}$, the label of the supercell $\mathcal{L}'_{(i,j)} = P(i, j)$.

Properties (1) and (2) define how sets of cells in P' replicate individual cells in P, and the labels of these sets of cells and individual cells. Property (3) restricts the region of each supercell not in the label region to contain only cells with a common fuzz label. Property (4) requires that each supercell's label matches the label of the corresponding cell in P.

4   SAS over PCFG Separation Lower Bound

This result uses a set of shapes we call n-*stagglers*, an example is seen in Figure 3. The shapes consist of $\log n$ bars of dimensions $n/\log n \times 1$ stacked vertically atop each other, with each bar horizontally offset from the bar below it by some amount in the range $-(n/\log n - 1), \ldots, n/\log n - 1$. We use the shorthand that $\log n = \lfloor \log n \rfloor$ for conciseness. Every sequence of $\log n - 1$ integers, each in the range $[-(n/\log n - 1), n/\log n - 1]$, encodes a unique staggler and by the pidgeonhole principle, some n-staggler requires $\log((2n/\log n - 1)^{\log n - 1} = \Omega(\log^2 n)$ bits to specify.

Lemma 1. *Any n-staggler can be derived by a PCFG of size $O(\log n)$.*

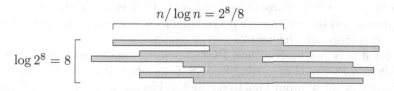

$$n/\log n = 2^8/8$$

$$\log 2^8 = 8$$

Fig. 3. The 2^8-staggler specified by the sequence $-18, 13, 9, -17, -4, 12, -10$

Proof. A set of $O(\log n)$ production rules deriving a bar (of size $\Theta(n/\log n) \times 1$) can be constructed by repeatedly doubling the length of the bar, using an additional $\log n$ rules to form the bar's exact length. The result of these production rules is a single non-terminal B deriving a complete bar.

Using the non-terminal B, a stack of k bars can be described using a production rule $N \to (B, (x_1, 0))(B, (x_2, 1)) \ldots (B, (x_k, k-1))$, where the x-coordinates $x_1, x_2, \ldots, x_k$ encode the offsets of each bar relative to the bar below it. An equivalent set of $k - 1$ production rules in binary normal form can be produced by creating a distinct non-terminal for T_i each stack of the first i bars, and a production rule $T_i \to (T_{i-1}, (0, 0))(B, (x_i, i))$ encoding the offset of the topmost bar relative to the stack of bars beneath it.

In total, $O(\log n)$ rules are used to create B, the non-terminal deriving a bar, and $O(\log n)$ are used to create the stack of bars, one per bar. So the n-staggler can be constructed using a PCFG of size $O(\log n)$.

Lemma 2. *For every n, there exists an n-staggler P such that any SAS or SSAS producing an assembly with label polyomino P has size $\Omega(\log^2 n / \log\log n)$.*

Proof. The proof is information-theoretic. Recall that more than half of all n-stagglers require $\Omega(\log^2 n)$ bits to specify. Now consider the number of bits contained in a SAS $\mathcal{S}$. Recall that $|\mathcal{S}|$ is the number of edges in the mix graph of $\mathcal{S}$. Any SAS can be encoded naively using $O(|\mathcal{S}| \log |\mathcal{S}|)$ bits to specify the mix graph, $O(|T| \log |T|)$ bits to specify the tile set, and $O(|\mathcal{S}| \log |T|)$ bits to specify the tile type at each leaf node of the mix graph. Because the number of tile types cannot exceed the size of the mix graph, $|T| \le |\mathcal{S}|$. So the total number of bits needed to specify $\mathcal{S}$ (and thus the number of bits of information contained in $\mathcal{S}$) is $O(|\mathcal{S}| \log |\mathcal{S}| + |T| \log |T| + |\mathcal{S}| \log |\mathcal{S}|) = O(|\mathcal{S}| \log |\mathcal{S}|)$. So some n-staggler requires a SAS $\mathcal{S}$ such that $O(|\mathcal{S}| \log |\mathcal{S}|) = \Omega(\log^2 n)$ and thus $|\mathcal{S}| = \Omega(\log^2 n / \log\log n)$.

Theorem 1. *The separation of SASs and SSASs over PCFGs is $\Omega(\log n / \log\log n)$.*

Proof. By the previous two lemmas, more than half of all n-stagglers require SASs and SSASs of size $\Omega(\log^2 n / \log\log n)$ and all n-stagglers have PCFGs of size $O(\log n)$. So the separation is $\Omega(\log n / \log\log n)$.

We also note that scaling the n-staggler by a c-factor produces a shape which is derivable by a CFG of size $O(\log n + \log c)$. That is, the result still holds for n-stagglers scaled by any amount polynomial in n. For instance, the $O(n)$-factor of the construction of Theorem 2.

At first it may not be clear how PCFGs achieve smaller encodings. After all, each rule in a PCFG G or mixing in SAS $\mathcal{S}$ specifies either a set of right-hand side symbols or set of input bins to use and so has up to $O(\log |G|)$ or $O(\log |\mathcal{S}|)$ bits of information. The key is the coordinate describing the location of each right-hand side symbol. These offsets have up to $O(\log n)$ bits of information and in the case that G is small, say $O(\log n)$, each rule has a number of bits *linear* in the size of the PCFG!

5 SAS over PCFG Separation upper Bound

Next we show that the separation lower bound of the last section is nearly large as possible by giving an algorithm for converting any PCFG G into a SSAS S with system size $O(|G| \log n)$ such that S produces an assembly that is a fuzzy replica of the polyomino derived by G. Before describing the full construction, we present approaches for efficiently constructing general binary counters and for simulating glues using geometry.

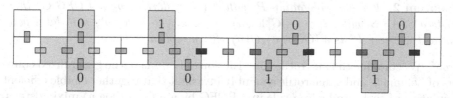

Increment 0011_b by 1, yielding 0100_b.

Fig. 4. A binary counter row constructed using single-bit constant-sized assemblies. Dark blue and green glues indicate 1-valued carry bits, light blue and green glues indicate 0-valued carry bits.

The *binary counter row assemblies* used here are a generalization of those by Demaine et al. [6] consisting of constant-sized bit assemblies, and an example is seen in Figure 4. Our construction achieves $O(\log n)$ construction of arbitrary ranges of rows and increment values, in contrast to the contruction of [6] that only produces row sets of the form $0, 1, \ldots, 2^{2^m} - 1$ that increment by 1. To do so, we show how to construct two special cases from which the generalization follows easily.

Lemma 3. *Let i, j, n be integers such that $0 \le i \le j < n$. There exists a SSAS of size $O(\log n)$ with a set of bins that, when mixed, assemble a set of $j - i + 1$ binary counter rows with values $i, i+1, \ldots, j$ incremented by 1.*

Lemma 4. *Let k, n be integers such that $0 \le k \le n$ and $n = 2^m$. There exists a SSAS of size $O(\log n)$ with a set of bins that, when mixed, assemble a set of 2^m binary counter rows with values $0, 1, \ldots, 2^m - 1$ incremented by k.*

Lemma 5. *Let i, j, k, n be integers such that $0 \le i \le j < n$ and $0 \le k \le n$. There exists a SSAS of size $O(\log n)$ with a set of bins that, when mixed, assemble a set of $j - i + 1$ binary counter rows with values $i, i+1, \ldots, j$ incremented by k.*

Proof. Combine the constructions used in the proofs of Lemmas 3 and 4 by using mixing sequences as in the proof of Lemma 3 and sets of four subassemblies encoding input, carry, and increment bit values as in the proof of Lemma 4.

Theorem 8 of Demaine et al. [6] describes how to reduce the number of glues used in a system by replacing each tile with a large *macrotile* assembly, and encoding the tile's glues via unique geometry on the macrotile's sides. We prove a similar result for labeled tiles, used for proving Theorems 2, 3, and 7.

Lemma 6. *Any mismatch-free* $\tau = 1$ *SAS (or SSAS)* $\mathcal{S} = (T, G, \tau, M)$ *can be simulated by a SAS (or SSAS)* $\mathcal{S}'$ *at* $\tau = 1$ *with* $O(1)$ *glues, system size* $O(\Sigma(T)|T| + |\mathcal{S}|)$, *and* $O(\log|G|)$ *scale.*

Armed with these tools, we are ready to convert PCFGs into SSASs. Recall that in Section 4 we showed that in the worst case, converting a PCFG into a SSAS (or SAS) *must* incur an $\Omega(\log n / \log \log n)$-factor increase in system size. Here we achieve a $O(\log n)$-factor increase.

Theorem 2. *For any polyomino* P *with* $|P| = n$ *derived by a PCFG* G, *there exists a SSAS* $\mathcal{S}$ *with* $|\mathcal{S}| = O(|G|\log n)$ *producing an assembly with label polyomino* P', *where* P' *is a* $(O(\log n), O(n))$-*fuzzy replica of* P.

Proof. We combine the macrotile construction of Lemma 6, the generalized counters of Lemma 5, and a macrotile assembly invariant that together enable efficient simulation of each production rule in a PCFG by a set of $O(\log n)$ mixing steps.

MACROTILES. The macrotiles used are extended versions of the macrotiles in Lemma 6 with two modifications: a secondary, *reservoir macroglue* assembly on each side of the tile in addition to a primary *bonding macroglue*, and a thin *cage* of dimensions $\Theta(n) \times \Theta(\log n)$ surrounding each reservoir macroglue (see Figure 5).

Mixing a macrotile with a set of bins containing counter row assemblies constructed by Lemma 5 causes completed (and incomplete) counter rows to attach to the macrotile's macroglues. Because each macroglue's geometry matches the geometry of exactly one counter row, a partially completed counter row that attaches can only be completed with bit assemblies that match the macroglue's value. As a result, mixing the bin sets of Lemma 5 with an assembly consisting of macrotiles produces the same set of products as mixing a completed set of binary counter rows with the assembly.

An attached counter row effectively causes the macroglue's value to change, as it presents geometry encoding a new value and covers the macroglue's previous value. The cage is constructed to have height sufficient to accomodate up to n counter rows attached to the reservoir macroglue, but no more.

Because of the cage, no two macrotiles can attach by their bonding macroglues unless the macroglue has more than n counter rows attached. Alternatively, one can produce a thickened counter row with thickness sufficient to extend beyond the cage. We call such an assembly a *macroglue activator*, as it "activates" a bonding macroglue to being able to attach to another promoted macroglue on another macrotile. Notice that a macroglue activator will never attach to a bonding macroglue's reservoir twin, as the cage is too small to contain the activator.

AN INVARIANT. Counter rows and activators allow precise control of two properties of a macrotile: the identities of the macroglues on each side, and whether these glues are activated. In a large assembly containing many macroglues, the ability to change and activate glues allows precise encoding of how an assembly can attach to others. In the remainder of the construction we maintain the invariant that every macrotile has the same glue identity on all four sides, and any

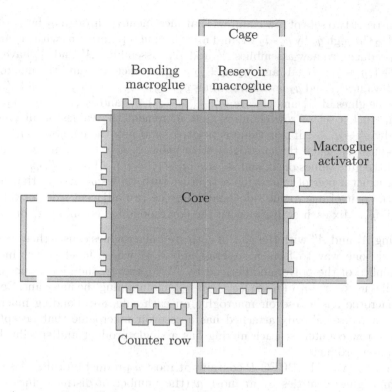

Fig. 5. A macrotile used in converting a PCFG to a SAS, and examples of value maintenance and offset preparation

macrotile assembly consists of macrotiles with glue identities forming a contiguous interval, e.g. 4, 5, 6, 7. Intervals are denoted $[i, i']$, e.g. $[g_4, g_7]$.

By Lemma 5, a set of row counters incrementing the glue identities of *all* glues on a macrotile can be produced using $O(\log n)$ work. Activators, by virtue of being nearly rectangular with $O(\log n)$ cells of bit geometry can also be produced using $O(\log n)$ work.

PRODUCTION RULE SIMULATION. Consider a PCFG with non-terminal N and production rule $N \rightarrow (R_1, (x_1, y_1))(R_2, (x_2, y_2))$ and a SSAS with two bins containing assemblies A_1, A_2 with the label polyominoes of A_1 and A_2 being fuzzy replicas of the polyominoes derived by R_1 and R_2. Also assume A_1 and A_2 are assembled from the macrotiles just described, including the invariant that the identities of the glues on A_1 and A_2 are identical on all sides of a macrotile and contiguous across the assembly, i.e. the identities of the glues are $[i_1, j_1]$ and $[i_2, j_2]$ on assemblies A_1 and A_2, respectively.

Select two cells c_{R_1}, c_{R_2}, in the polyominoes derived by R_1 and R_2 adjacent in polyomino derived by N. Define the glue identities of the two macrotiles forming the supercells mapped to c_{R_1} and c_{R_2} to be g_1 and g_2. Then the glue sets on A_1 and A_2 can be decomposed into three subsets $[i_1, g_1 - 1]$, $[g_1]$, $[g_1 + 1, j_1]$ and $[i_2, g_2 - 1]$, $[g_2]$, $[g_2 + 1, j_2]$, respectively. We change these glue values in three steps:

1. Construct two sets of row counters that increment i_1 through g_1 by $j_1 - i_1 + 1$ and i_2 through g_2 by $g_2 - i_2 + 1$, and mix them in separate bins with A_1 and A_2 to produce two new assemblies A_1' and A_2'. Assemblies A_1' and A_2' have glues $[g_1 + 1, g_1 + j_1 - i_1 + 1]$ and $[g_2, g_2 + j_2 - i_2]$, respectively, and the macroglues with values g_1 and g_2 now have values $g_1' = g_1 + (g_1 - i_1) + j_1 + 1$ and $g_2' = g_2$, i.e. the glues of A_1' and A_2' are $[g_1' - (j_1 - i_1), g_1']$ and $[g_2', g_2' + j_2 - i_2]$.
2. Construct a set of row counters that increment the values of all glues on A_2' by $g_2' - g_1' + 1$ if this value is positive, and mix the counters with A_2' to produce A_2''. Then the macroglue with value g_2' now has value $g_2'' = g_1' + 1$ and the glue values of A_1' and A_2'' are $[g_1' - (j_1 - i_1), g_1']$ and $[g_2'', g_2'' + j_2 - i_2]$.
3. Construct a pair of macroglue activators with values g_1' and g_2'' that attach to the pair of macroglue sides matching the two adjacent sides of cells c_{R_1} and c_{R_2}. Mix each activator with the corresponding assembly A_1' or A_2''.

Mixing A_1' and A_2'' with the pair of activated macroglues causes them to bond in exactly one way to form a superassembly A_3 whose label polyomino is a fuzzy replica of the polyomino derived by N. Moreover, the glue values of the macrotiles in A_3 are $[g_1' - (j_1 - i_1), g_2'' + j_2 - i_2]$, maintaining the invariant. Because each macrotile has a reservoir macroglue on each side, any bonding macroglue with an activator already attached has a reservoir macroglue that accepts the matching row counter, so each mixing has a single product and specifically no row counter products.

SYSTEM SCALE The PCFG P contains at most n production rules. Also, each step shifts glue identities by at most n (the number of distinct glues on the macrotile), so the largest glue identity on the final macrotile assembly is n^2. So we produce macrotiles with core assemblies of size $O(\log n) \times O(\log n)$ and cages of size $O(n)$. Assembling the core assemblies, cages, and initial macroglue assemblies of the macrotiles takes $O(|P| \log n + \log n + \log n) = O(|P| \log n)$ work, dominated by the core assembly production. Simulating each production rule of the grammar takes $O(\log n)$ work spread across a constant number of $O(\log n)$-sized sequences of mixings to produce sets of row counters and macroglue activators.

Applying Lemma 6 to the construction (creating macrotiles of macrotiles) gives a constant-glue version of Theorem 2:

Theorem 3. *For any polyomino P with $|P| = n$ derived by a PCFG G, there exists a SSAS S' using $O(1)$ glues with $|S'| = O(|G| \log n)$ producing an assembly with label polyomino P', where P' is a $(O(\log n \log \log n), O(n \log \log n))$-fuzzy replica of P.*

Proof. The construction of Theorem 2 uses $O(\log n)$ glues, namely for the counter row subconstruction of Lemma 5. With the exception of the core assemblies, all tiles of S have a common fuzz (gray) label, so creating macrotile versions of these tiles and carrying out all mixings involving these macrotiles and *completed* core assemblies is possible with $O(1 \cdot |T| + |S|) = O(|S|)$ mixings and scale $O(\log \log n)$. Scaled core assemblies of size $\Theta(n \log \log n) \times \Theta(n \log \log n)$ can be constructed using constant glues and $O(\log(n \log \log n)) = O(\log n)$ mixings, the same number of mixings as the unscaled $\Theta(n) \times \Theta(n)$ core assemblies of Theorem 2.

So in total, this modified construction has system size $O(|S|) = O(|G| \log n)$ and scale $O(\log \log n)$. Thus it produces an assembly with label polyomino that is a $(O(\log n \log \log n), O(n \log \log n))$-fuzzy replica of P.

The results in this section and Section 4 achieve a "one-sided" correspondence between the smallest PCFG and SSAS encoding a polyomino, i.e. the smallest PCFG is approximately an *upper bound* for the smallest SSAS (or SAS). Since the separation upper bound proof (Theorem 2) is constructive, the bound also yields an algorithm for converting a a PCFG into a SSAS.

6   PCFG over SAS and SSAS Separation Lower Bound

Here we develop a sequence of PCFGs over SAS and SSAS separation results, all within a polylogarithmic factor of optimal. The results also hold for polynomially scaled versions of the polyominoes, which is used to prove Theorem 7 at the end of the section. This scale invariance also surpasses the scaling of the fuzzy replicas in Theorems 2 and 3, implying that this relaxation of the problem statement in these theorems was not unfair.

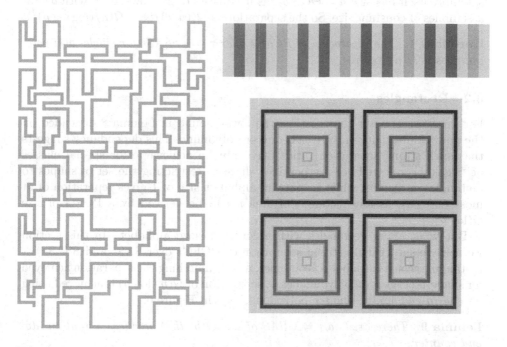

Fig. 6. Two-bit examples of the weak (left), end-to-end (upper right), and block (lower right) binary counters used to achieve separation of PCFGs over SASs and SSASs in Section 6

6.1 General Shapes

We show that the separation of PCFGs over SASs and SSASs is $\Omega(n/\log n)$ using a *weak binary counter*, seen in Figure 6. These shapes are macrotile versions of the doubly-exponential counters found in [6] with three modifications:

1. Each row is a single path of tiles, and any path through an entire row uniquely identifies the row.
2. Adjacent rows do not have adjacent pairs of tiles, i.e. they do not touch.
3. Consecutive rows attach at alternating (east, west, east, etc.) ends.

Lemma 7. *There exists a $\tau = 1$ SAS of size $O(b)$ that produces a 2^b-bit weak counter.*

Lemma 8. *For any PCFG G deriving a 2^b-bit weak counter, $|G| = \Omega(2^{2^b})$.*

Theorem 4. *The separation of PCFGs over $\tau = 1$ SASs for single-label polyomines is $\Omega(n/(\log\log n)^2)$.*

Proof. By the previous two lemmas, there exists a SAS of size $O(b)$ producing a b-bit weak counter, and any PCFG deriving this shape has size $\Omega(2^{2^b})$. The assembly itself has size $n = \Theta(2^{2^b} b)$, as it consists of 2^{2^b} rows, each with b sub-assemblies of constant size. So the separation is $\Omega((n/b)/b) = \Omega(n/(\log\log n)^2)$.

Corollary 1. *The separation of PCFGs over $\tau = 1$ SSASs for single-label polyominoes is $\Omega(n/\log^2 n)$.*

6.2 Rectangles

For the weak counter construction, the lower bound in Lemma 8 depended on the poor connectivity of the weak counter polyomino. This dependancy suggests that such strong separation ratios may only be achievable for special classes of "weakly connected" or "serpentine" shapes. Restricting the set of shapes to rectangles or squares while keeping an alphabet size of 1 gives separation of at most $O(\log n)$, as any rectangle of area n can be derived by a PCFG of size $O(\log n)$.

But what about rectangles with a constant-sized alphabet? In this section we achieve surprisingly strong separation of PCFGs over SASs and SSASs for rectangular constant-label polyominoes, nearly matching the separation achieved for single-label general polyominoes. A separation of $\Omega(n/\log n)$ is achieved using an *end-to-end binary counter* polyomino, seen in Figure 6.

Lemma 9. *There exists a $\tau = 1$ SAS of size $O(b)$ that produces a b-bit end-to-end counter.*

Lemma 10. *For any PCFG G deriving a b-bit end-to-end counter, $|G| = \Omega(2^b)$.*

Theorem 5. *The separation of PCFGs over $\tau = 1$ SASs for constant-label rectangles is $\Omega(n/\log^3 n)$.*

6.3   Squares

The rectangular polyomino of the last section has exponential aspect ratio, suggesting that this shape requires a large PCFG because it approximates a patterned one-dimensional assemblies reminiscent of those in [7]. Creating a polyomino with better aspect ratio but significant separation is possible by extending the polyomino's labels vertically. For a square this approach gives a separation of PCFGs over SASs of $\Omega(\sqrt{n}/\log n)$, non-trivial but far worse than the rectangle.

Our final result achieves $\Omega(n/\log n)$ separation of PCFGs over SASs for squares using a *block binary counter* (seen in Figure 6). Each "row" of the counter is actually a set of concentric square rings called a *block*.

Lemma 11. *For even b, there exists a $\tau = 1$ SAS of size $O(b)$ that produces a b-bit block counter.*

Lemma 12. *For any PCFG G deriving a b-bit block counter, $|G| = \Omega(2^b)$.*

Theorem 6. *The separation of PCFGs over $\tau = 1$ SASs for constant-label squares is $\Omega(n/\log^3 n)$.*

6.4   Constant-Glue Constructions

Lemma 6 proved that any system $\mathcal{S}$ can be converted to a slightly larger system (both in system size and scale) that simulates $\mathcal{S}$. Applying this lemma to the constructions of Section 6 yields identical results for constant-glue systems:

Theorem 7. *All results in Section 6 hold for systems with $O(1)$ glues.*

7   Conclusion

As the results of this work show, efficient staged assembly systems may use a number of techniques including, but not limited to, those described by local combination of subassemblies as captured by PCFGs. It remains an open problem to understand how the efficient assembly techniques of Section 5 and Section 6 relate to the general problem of optimally assembling arbitrary shapes.

Acknowledgements. We thank Benjamin Hescott and anonymous reviewers for helpful comments and feedback that greatly improved the presentation of the paper.

References

1. Adleman, L., Cheng, Q., Goel, A., Huang, M.-D.: Running time and program size for self-assembled squares. In: Proceedings of Symposium on Theory of Computing, STOC (2001)

2. Cannon, S., Demaine, E.D., Demaine, M.L., Eisenstat, S., Patitz, M.J., Schweller, R.T., Summers, S.M., Winslow, A.: Two hands are better than one (up to constant factors): Self-assembly in the 2HAM vs. aTAM. In: Proceedings of International Symposium on Theoretical Aspects of Computer Science (STACS). LIPIcs, vol. 20, pp. 172–184 (2013)
3. Charikar, M., Lehman, E., Lehman, A., Liu, D., Panigrahy, R., Prabhakaran, M., Sahai, A., Shelat, A.: The smallest grammar problem. IEEE Transactions on Information Theory 51(7), 2554–2576 (2005)
4. Chen, H.L., Doty, D.: Parallelism and time in hierarchical self-assembly. In: Proceedings of ACM-SIAM Symposium on Discrete Algorithms, SODA (2012)
5. Cook, M., Fu, Y., Schweller, R.: Temperature 1 self-assembly: determinstic assembly in 3D and probabilistic assembly in 2D. In: Proceedings of ACM-SIAM Symposium on Discrete Algorithms (SODA) (2011)
6. Demaine, E.D., Demaine, M.L., Fekete, S., Ishaque, M., Rafalin, E., Schweller, R., Souvaine, D.: Staged self-assembly: nanomanufacture of arbitrary shapes with $O(1)$ glues. Natural Computing 7(3), 347–370 (2008)
7. Demaine, E.D., Eisenstat, S., Ishaque, M., Winslow, A.: One-dimensional staged self-assembly. Natural Computing (2012)
8. Doty, D., Lutz, J.H., Patitz, M.J., Schweller, R.T., Summers, S.M., Woods, D.: Intrinsic universality in self-assembly. In: Proceedings of Symposium on Theoretical Aspects of Computer Science (STACS). LIPIcs, vol. 5, pp. 275–286 (2010)
9. Doty, D., Lutz, J.H., Patitz, M.J., Schweller, R.T., Summers, S.M., Woods, D.: The tile assembly model is intrinsically universal. In: Proceedings of Foundations of Computer Science (FOCS), pp. 302–310 (2012)
10. Jeż, A.: Approximation of grammar-based compression via recompression. Technical report, arXiv (2013)
11. Lehman, E.: Approximation Algorithms for Grammar-Based Data Compression. PhD thesis. MIT (2002)
12. Rothemund, P.W.K., Winfree, E.: The program-size complexity of self-assembled squares. In: Proceedings of Symposium on Theory of Computing (STOC), pp. 459–468 (2000)
13. Soloveichik, D., Winfree, E.: Complexity of self-assembled shapes. In: Ferretti, C., Mauri, G., Zandron, C. (eds.) DNA 11. LNCS, vol. 3384, pp. 344–354. Springer, Heidelberg (2005)
14. Winfree, E.: Algorithmic Self-Assembly of DNA. PhD thesis, Caltech (1998)

Functional Analysis of Large-Scale DNA Strand Displacement Circuits

Boyan Yordanov, Christoph M. Wintersteiger, Youssef Hamadi,
Andrew Phillips, and Hillel Kugler

Microsoft Research, Cambridge UK
{yordanov,cwinter,youssefh,aphillip,hkugler}@microsoft.com
http://research.microsoft.com/z3-4biology

Abstract. We present a method for the analysis of functional properties of large-scale DNA strand displacement (DSD) circuits based on Satisfiability Modulo Theories that enables us to prove the functional correctness of DNA circuit designs for arbitrary inputs, and provides significantly improved scalability and expressivity over existing methods. We implement this method as an extension to the Visual DSD tool, and use it to formalize the behavior of a 4-bit square root circuit, together with the components used for its construction. We show that our method successfully verifies that certain designs function as required and identifies erroneous computations in others, even when millions of copies of a circuit are interacting with each other in parallel. Our method is also applicable in the verification of properties for more general chemical reaction networks.

1 Introduction

The engineering of nanoscale devices from DNA has emerged as a powerful technology, with potential applications in nanomedicine and nanomaterials. More recently, DNA strand displacement (DSD) has attracted attention as a promising approach for engineering molecular devices with complex dynamics [24], and has been shown to scale to large circuits [17]. In spite of this potential, many challenges remain before the design of DSD circuits with predictable, robust behavior becomes routine. In addition to the experimental difficulties of synthesis, assembly, and elimination of cross-talk, the massive parallelism and complexity of DSD circuits make their manual design challenging and error-prone.

A number of computational methods and tools have been developed to facilitate the design process. In particular, the Visual DSD tool [14] computes the set of all possible strand displacement reactions generated from an initial collection of DNA species, and simulates these reactions over time. Methods have also been developed for proving that a set of strand displacement reactions is equivalent to a reduced set of reactions [18,7]. However, further work is needed to be able to state and prove properties about the function that these reactions perform. To help address this, methods based on probabilistic model checking have been developed to prove properties about the states that a strand displacement circuit

D. Soloveichik and B. Yurke (Eds.): DNA 2013, LNCS 8141, pp. 189–203, 2013.
© Springer International Publishing Switzerland 2013

traverses, together with the expected time and probability of failure [13]. So far however, these methods do not scale to realistic numbers of molecules. To help improve scalability, a symbolic method called Z34Bio for analyzing large and potentially infinite state spaces based on Satisfiability Modulo Theories (SMT) was developed [21,22]. This technique has been applied to study structural properties of DNA circuits, such as the presence of exposed DNA sequences.

In this paper we present a method that allows the desired properties of a DNA strand displacement circuit to be formalized as a high-level functional specification, and formally verified for realistic numbers of molecules. The method extends the use of SMT-solvers for analyzing chemical reaction networks presented in [21], and is implemented within the Visual DSD tool using the Z34Bio framework, which is based on the Z3 theorem prover and SMT-solver [6]. To illustrate this approach, we study a model of the 4-bit square root circuit described in [2], which was originally developed as a localized circuit in contrast to the design from [17]. We formalize and analyze functional properties of the individual components used to construct this system, and show that a modified version of this design functions correctly, even when millions of copies interact with each other in parallel. We illustrate how our method helps identify design errors at both the component and circuit level. Although the method has been tailored specifically for DNA strand displacement systems, it is also more generally applicable for the analysis of chemical reaction networks.

2   SMT Analysis of Chemical Reaction Networks

This section summarizes the SMT-based method for analyzing Chemical Reaction Networks (CRNs) presented in [21], which will be used in the remainder of the paper. We denote a finite set as $S = \{s_0, \ldots, s_N\}$, where $|S| = N + 1$ is the number of elements in S. We use $S = \{(s_0, n_0), \ldots, (s_N, n_N)\}$ to denote a finite multiset where each pair (s_i, n_i) denotes an element s_i and its multiplicity n_i, with $n_i > 0$. Given a multiset S we use $s \in S$ for $\exists n . (s, n) \in S$ and $S(s) = n$ when $(s, n) \in S$ and $S(s) = 0$ otherwise. We define a CRN as a pair $(\mathcal{S}, \mathcal{R})$, where $\mathcal{S}$ is a finite set of species and $\mathcal{R}$ is a finite set of possible reactions. A reaction $r \in \mathcal{R}$ is defined as a pair of multisets $r = (R_r, P_r)$ denoting the reactants and products of r, respectively. For $(s, n) \in R_r$ (respectively P_r), $s \in \mathcal{S}$ is a species and n is the stoichiometry indicating how many molecules of s are consumed (respectively produced) when reaction r takes place.

To study the dynamics of a CRN with single-molecule resolution, we formalize its behavior as the transition system $\mathcal{T} = (Q, q_0, T)$, where Q is the set of states, $q_0 \in Q$ is the uniquely defined initial state, and $T \subseteq Q \times Q$ is the transition relation. Each state $q \in Q$ is a multiset of species and $q(s)$ indicates how many molecules of s are present in state q. A reaction r is *enabled* in q if there are enough molecules of each of its reactants for it to trigger; i.e., $enabled(r, q) \leftrightarrow \bigwedge_{s \in \mathcal{S}} q(s) \geq R_r(s)$. A state q is *terminal* if it has no enabled reactions i.e. $terminal(q) \leftrightarrow \bigwedge_{r \in \mathcal{R}} \neg enabled(r, q)$. The transition relation T is defined as

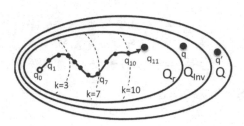

Fig. 1. SMT-based Analysis. All trajectories explored by Bounded Model Checking represent valid computations from q_0 but Q_r^k (*e.g.* $k = 3, 7, 10$) generally under-approximates the reachable states Q_r. The over-approximation Q_{Inv} allows proving that state q' is unreachable but spurious states (*e.g.* q) may be included.

$$T(q, q') \leftrightarrow \bigvee_{r \in \mathcal{R}} (enabled(r, q) \land \bigwedge_{s \in S} q'(s) = q(s) - R_r(s) + P_r(s)). \qquad (1)$$

Once a CRN $(\mathcal{S}, \mathcal{R})$ is encoded as a transition system $\mathcal{T} = (Q, q_0, T)$ with some initial state $q_0 \in Q$, a number of analysis questions are expressible as logical formulas and resolvable using model checking methods and SMT solvers such as Z3 [6]. In the following, we focus on safety properties [16] *i.e.* a given state predicate $P : Q \to \mathbb{B}$ holds for all reachable states. Let $Q_r \subseteq Q$ denote all reachable states of $\mathcal{T}$ and let $Q_r^k \subseteq Q$ denote the set of states reachable in up to k transitions from q_0 (note that $Q_r^k \subseteq Q_r^{k+1} \subseteq \cdots \subseteq Q_r$).

We check the reachability of a state $q \in Q$ such that $\neg P(q)$ holds by using an SMT solver to decide whether the formula $\exists q_1, \ldots, q_k \cdot \bigwedge_{i=0}^{k-1} T(q_i, q_{i+1}) \land \bigvee_{i=0}^{k} \neg P(q_i)$ is satisfiable. The formula represents the *unrolling* of the transition relation for k transitions from q_0 (see Fig. 1 for an example). This type of approach is usually called Bounded Model Checking (BMC) [1] and is particularly good at finding bugs in hardware and software, with the added advantage of producing an explicit computation trace which demonstrates the behavior that leads to the violation of P. However, since only bounded executions are considered and Q_r^k generally under-approximates Q_r for feasible choices of k (see [4] for a discussion on the length of computation traces), this technique only serves to prove that no errors are encountered in a finite number of transitions.

As a complementary approach based on inductive invariants, a state invariant Q_{Inv} such that $Q_r \subseteq Q_{Inv} \subseteq Q$, allows us to prove that all reachable states satisfy P because

$$\nexists q \in Q_{Inv} \cdot \neg P(q) \; \to \; \nexists q \in Q_r \cdot \neg P(q) \; \to \; \forall q \in Q_r \cdot P(q).$$

However, due to the over-approximation, this is not sufficient to prove the existence of reachable states violating the given property (see Fig. 1), since we have

$$\exists q \in Q_{Inv} \cdot \neg P(q) \; \nrightarrow \; \exists q \in Q_r \cdot \neg P(q).$$

An invariant Q_{Inv} is computable through strategies developed for the analysis of Petri nets [11], metabolic networks [9], and DNA circuits [21], where it captures constraints such as mass-conservation (various techniques from hardware and software analysis, *e.g.*, abstract interpretation [5] also apply).

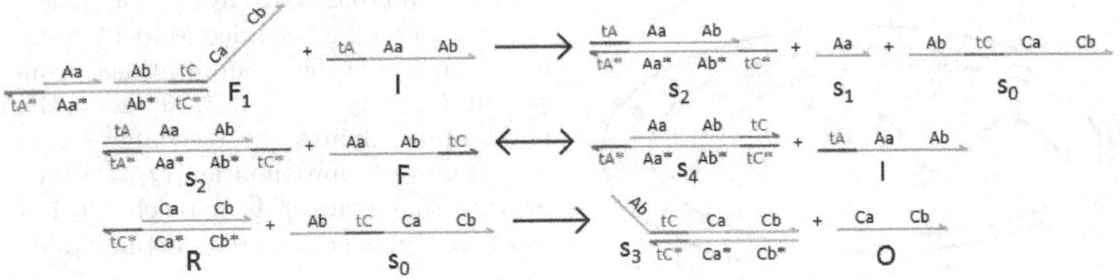

Fig. 2. FANOUT Gate

To analyze chemical reaction networks and DNA circuits, we combine the BMC and state invariant approaches, where we prove that states with given properties are unreachable in any number of steps using Q_{Inv} but use BMC to guarantee the reachability of states and identify finite computation traces with specific behavior.

3   SMT Analysis of DNA Strand Displacement Circuits

In the following, we focus on circuits constructed as DNA strand displacement (DSD) devices [24], which is the DNA computing paradigm supported in Visual DSD [14]. In [21] we utilized the known structure of DNA species in these systems to develop an approach for the computation of constraints that was specific to DSD circuits. Intuitively, the individual DNA strands from which all species in the system are composed are preserved and their total amounts remain unchanged. A set of constraints was computed to capture this *conservation of strands* property, which allowed the computation of a state invariant Q_{Inv} (Sec. 2) based on the numbers of strands present initially (in state q_0). In Fig. 3 we illustrate this computation for the FANOUT gate shown in Fig. 2 (a component of the circuit we study in Sec. 4), which produces multiple copies of the output species O through a reporter R for a given input I, where the degree of "fanout" is controlled through the amount of species F.

Throughout the rest of this section, we present new strategies extending the approach and enabling the application of SMT-based methods to the analysis of large scale DSD circuits.

3.1   Identification of Inactive Reactions

As discussed in Sec. 2, our analysis strategy based on state invariants (*e.g.* computed using the method from [21] as in Fig. 3) is conservative, leading to the possible identification of spurious (unreachable) states. We exemplify this[1] on the FANOUT gate (Fig. 2), for which the following constraints are derived when the system is initialized in state $q_0 = \{(R, 10), (F_1, 2), (F, 1)\}$:

---

[1] Note that this example is similar to the one from Fig. 3, with the exception of the input $q_0(I) = 0$.

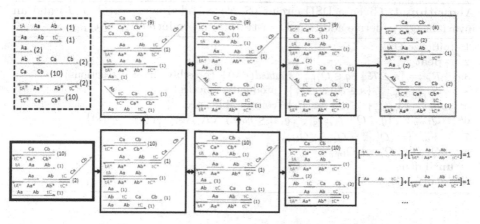

Fig. 3. Strand conservation along a computation of the FANOUT gate (Fig. 2). The system is initialized in $q_0 = \{(R, 10), (I, 1), (F_1, 2), (F, 1)\}$ (thick black border) and terminates in the state shown with a red border. The total number of DNA strands (top left box) remains unchanged in each state, which is captured using a set of constraints (*e.g.* bottom right). Each constraint captures information about a DNA strand shared between species (*i.e.* in each state, the sum of the molecule numbers of all species sharing strand s is equal to the number of s present in the system), which defines a state invariant Q_{Inv}. See Eqn.(2) for a related example and [21] for additional details.

$$q \in Q_{Inv} \leftrightarrow q \in Q \land$$
$$[q(s_2) + q(I) = 0] \land [q(s_1) + q(F_1) = 2] \land [q(s_4) + q(F) = 1] \land$$
$$[q(s_3) + q(s_0) + q(F_1) = 2] \land [q(O) + q(R) = 10] \land \tag{2}$$
$$[q(s_4) + q(s_2) + q(F_1) = 2] \land [q(s_3) + q(R) = 10] .$$

Given these constraints, only two terminal states $q_1, q_2 \in Q_{Inv}$ are possible where $q_1 = q_0$ and

$$q_2 = \{(R, 9), (O, 1), (F_1, 1), (s_1, 1), (s_3, 1), (s_4, 1)\} .$$

While q_1 satisfies the expected behavior of the circuit (no output is produced without input) state q_2 violates it. However, due to the over-approximation of Q_{Inv} (see Sec. 2), this does not directly imply that the FANOUT gate is flawed.

A closer inspection of this example reveals that, when initialized in state q_0, no reactions are enabled for this system but such information is not captured in the derived constraints and therefore spurious terminal states are identified. To decrease this conservativeness, we use the available constraints Q_{Inv} to identify reactions that are disabled for any reachable state of the system (this might require a call to the SMT solver for each $r \in \mathcal{R}$ but does not involve deep reasoning for most). Then, we use this information to identify species which are never produced (resp. consumed) and constrain their abundances to only decrease (resp. increase) from their initial values. The procedure is repeated iteratively until no additional constraints are derived (see Alg. 1).

Algorithm 1. Given a DSD circuit $(\mathcal{S}, \mathcal{R})$ encoded as $\mathcal{T} = (Q, q_0, T)$ and an invariant Q_{Inv}, derive additional constraints to produce $Q'_{Inv} \subseteq Q_{Inv}$

1: Initialize $Q'_{Inv} := Q_{Inv}$
2: **repeat**
3:     $\mathcal{R}_e := \{r \in \mathcal{R} \mid \exists q \in Q'_{Inv} \cdot enabled(r,q)\}$ {possibly enabled reactions}
4:     $\mathcal{S}_p := \bigcup_{r \in \mathcal{R}_e} \{s \in \mathcal{S} \mid s \in P_r\}$ {producible species}
5:     $\mathcal{S}_c := \bigcup_{r \in \mathcal{R}_e} \{s \in \mathcal{S} \mid s \in R_r\}$ {consumable species}
6:     $Q_{tmp} := \{q \in Q \mid \bigwedge_{s \in (\mathcal{S} \setminus \mathcal{S}_p)} q(s) \leq q_0(s) \wedge \bigwedge_{s \in (\mathcal{S} \setminus \mathcal{S}_c)} q(s) \geq q_0(s)\}$
7:     done := $(Q'_{Inv} = Q'_{Inv} \setminus Q_{tmp})$
8:     $Q'_{Inv} := Q'_{Inv} \setminus Q_{tmp}$
9: **until** done
10: **return** Q'_{Inv}

Applying Alg. 1 to the FANOUT gate produces the following additional constraints, which are sufficient to eliminate the spurious terminal state q_2:

$$q \in Q'_{Inv} \leftrightarrow q \in Q_{Inv} \wedge$$
$$[q(F) \geq 1] \wedge [q(R) \leq 10] \wedge [q(I) \leq 0] \wedge [q(F) \leq 1] \wedge [q(s_2) \leq 0] \wedge$$
$$[q(s_1) \leq 0] \wedge [q(s_0) \leq 0] \wedge [q(s_4) \leq 0] \wedge [q(F_1) \leq 2] \wedge [q(F_1) \geq 2] .$$

The use of Alg. 1 is not guaranteed to eliminate all unreachable states captured in Q_{Inv}. In other words, even though $Q'_{Inv} \subseteq Q_{Inv}$, the invariant still over-approximates the reachable states (*i.e.* in general, $Q_r \subseteq Q'_{Inv}$) and, therefore, unreachable states $q \notin Q_r$ such that $q \in Q'_{Inv}$ might still exist. Even so, the invariant strengthening strategy implemented in Alg. 1 is useful, particularly for the analysis of DNA circuits as the ones discussed in Sec. 4. For such designs, system inputs are encoded using the availability of chemical species where, for specific input values, certain species are not supplied. In these cases, Alg. 1 identifies reactions that are never enabled and restricts Q'_{Inv} accordingly.

3.2 Encoding Generalization

To identify erroneous computations for large DSD circuits such as the ones from Sec. 3, a BMC strategy requires prohibitively long paths, since the transition relation from Eqn. (1) only captures the execution of a single reaction per step. In the following, we relax this requirement by abstracting the exact number of consecutive executions of a reaction. Given states $q, q' \in Q$

$$\text{reach}(q, q', r, n) = \left[\begin{array}{cc} 0 \leq n \leq min_{s \in R_r}\{\lfloor \frac{q(s)}{R_r(s)} \rfloor\} & \wedge \\ \bigwedge_{s \in S} q'(s) = q(s) - nR_r(s) + nP_r(s) & \end{array} \right]$$

expresses the property that state q' is reachable from q through n consecutive executions of reaction $r \in \mathcal{R}$. The condition that the reaction is enabled is implicitly captured in the choice of n (*i.e.* if r is disabled, then $0 \leq n \leq 0$). The transition relation

$$T(q, q') \leftrightarrow q' \in Q_{Inv} \wedge \bigvee_{r \in \mathcal{R}} \exists n \,.\, \text{reach}(q, q', r, n)$$

captures multiple executions of the same reaction in a given step and therefore allows us to consider shorter computation traces. Note that this does not influence the completeness of the approach as single-reaction steps are also allowed. Furthermore, available constraints (*e.g.* derived as in Sec. 3.1) are captured directly in the transition relation.

Besides re-defining the transition relation, we generalize the transition system representation of a DSD circuit to $\mathcal{T} = (T, Q_0, Q)$ where $Q_0 \subseteq Q$ is a (possibly infinite) set of initial states. While currently, a circuit is defined using a unique initial state (population of species) within Visual DSD, it is natural to reason about the behavior of certain systems under a range of possible inputs, encoded through the abundances of chemical species at the beginning of a computation, which we illustrate in Sec. 4. Note that an invariant computed as in [21] depends on the initial state and is therefore denoted as $Q_{Inv}(q), q \in Q_0$ in the following. While applying Alg. 1 to explicit initial states $q_0 \in Q_0$ was sufficient for the circuits we consider in Sec. 3, additional strategies are required for other systems.

3.3 Implementation of Methods in Visual DSD

We developed a prototype implementation of the methods reviewed in Sec. 2, together with the extension described in this section, as part of Visual DSD. The implementation makes us of the Z3 theorem prover [6], which provides efficient decision procedures for several theories including bit-vectors [20]. This allows us to specify various system properties and automatically verify them during the DSD circuit design process. The experimental results presented in Sec. 4 were obtained on 2.5 GHz Intel L5420 CPU machines with a 2 GB memory limit where computation required under a minute per benchmark.

4 Functional Analysis of a 4-bit Square Root Circuit

In this section, we study the 4-bit square root circuit design from [2]. First, we formalize and analyze the functional behavior of the individual components used for the construction of this system and then apply our method to study properties of the full system, when multiple copies of the circuit are operating in parallel. For each component, we define a set of initial states $Q_0 \subset Q$ capturing the possible abundances of species (inputs, gates, etc) present at the beginning of a computation and a property $P(q_0, q)$ that describes the expected output for a given input, encoded as part of the initial state q_0. To prove the correctness of a circuit, we need to show that $\forall q_0 \in Q_0, q \in Q_r \,.\, terminal(q) \rightarrow P(q_0, q)$ (*i.e.* for all input values, the correct output is produced when the computation terminates, regardless of the initial abundances of other species). We compute

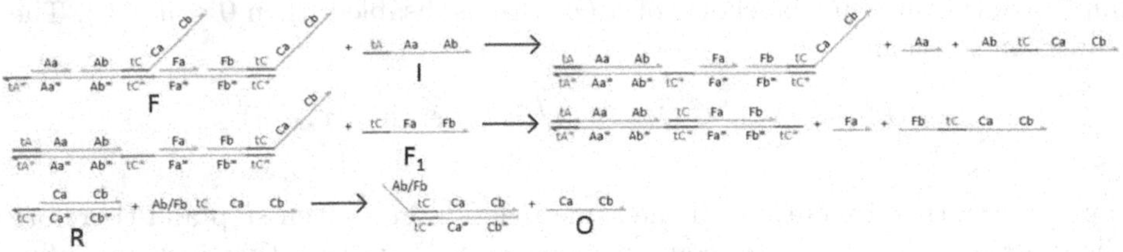

Fig. 4. Modified `FANOUT` Gate with fanout degree of two

the state invariant Q_{Inv} using the procedure from [21] illustrated in Fig. 3 (or Q'_{Inv} extended through Alg. 1) and use it to prove that $terminal(q) \wedge \neg P(q_0, q)$ is not satisfiable for states $q_0 \in Q_0$ and $q \in Q_{Inv}(q_0)$ (*i.e.* no terminal state exists where incorrect output is produced). This formula is trivially unsatisfiable when no terminal states exist (*i.e.* $\forall q \in Q_{Inv} . \neg terminal(q)$) and, to conclude the proof, we show that this is not the case. When incorrect behavior is identified through this strategy (as is the case for one of the square root circuit designs we explore), we use BMC (see Sec. 2) to identify an error trace.

FANOUT Gate. The `FANOUT` gate (Fig. 2) introduced in Sec. 3.1 is intended to split a particular input species I into multiple copies of output O, where the degree of fanout is controlled through species F. We define the set $Q_0 = \{q \in Q'_{Inv} \mid q(s) > 0 \text{ if } s \in \{R, F_1\} \text{ and } q(s) = 0 \text{ if } s \notin \{R, F_1, I, F\}\}$ (*i.e* gates F_1, reporters R and possibly fanout F and input I species are present initially). We used our analysis approach to prove that, for all initial states $q_0 \in Q_0$ and terminal states $q \in Q'_{Inv}(q_0)$ the behavior of the component formalized as

$$P_{\text{FANOUT}}(q_0, q) \leftrightarrow q(O) = \begin{cases} q_0(I) + q_0(F) & \text{when } q_0(I) > 0 \\ 0 & \text{otherwise} \end{cases}$$

holds, as long as the additional conditions $q_0(R) \geq q_0(I) + q_0(F)$ and $q_0(F_1) \geq I$ are satisfied (*i.e.* there is an excess of reporters and gates). Note that this behavior holds regardless of the specific input $q_0(I)$ and fanout $q_0(F)$ settings. Thus, the component adds a constant to the input value (when input is present) but replicates the desired behavior $q(O) = m \cdot q_0(I)$ only when $q_0(F) = (m - 1) \cdot q_0(I)$ and, as a result, $q_0(F)$ must be precisely tuned for a specific input value. To obtain the correct behavior for arbitrary inputs, we redesign the gate for fanout $m = 2$ as in Fig. 4 and show that the expected behavior

$$P_{\text{FANOUT2}}(q_0, q) \leftrightarrow q(O) = 2 \cdot q_0(I)$$

is now satisfied when $q_0(F) \geq q_0(I)$, $q_0(R) \geq 2 \cdot q_0(I)$, and $q_0(F_1) \geq 2 \cdot q_0(F)$ (*i.e.* gates and reporters are in excess).

AND Gate. The `AND` gate (Fig. 5) is a component designed to implement the corresponding logical operation. We define the set $Q_0 = \{q \in Q_{Inv} \mid q(s) >$

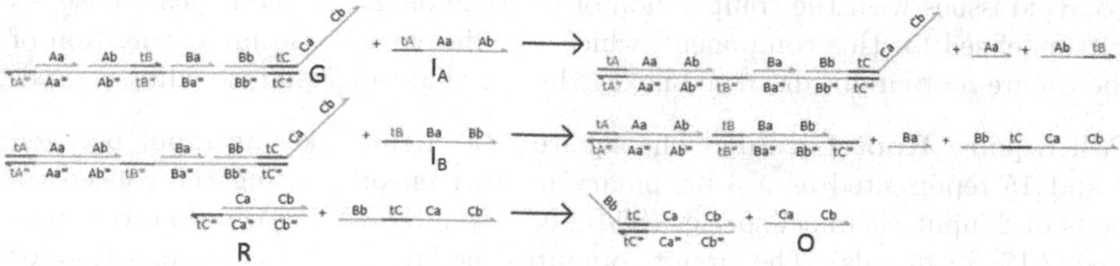

Fig. 5. AND Gate

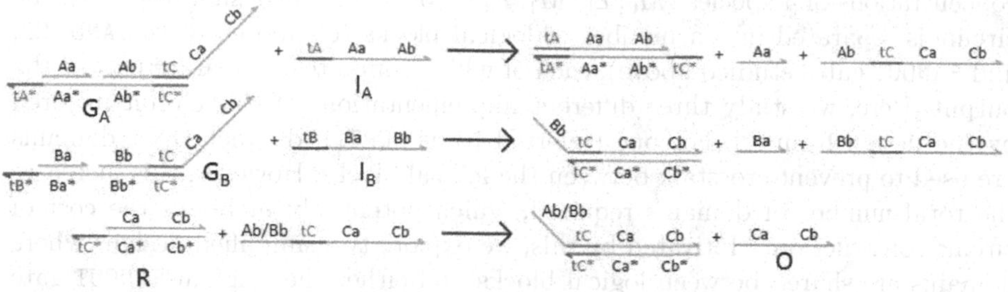

Fig. 6. OR Gate

0 if $s \in \{R, G\}$ and $q(s) = 0$ if $s \notin \{R, G, I_A, I_B\}\}$ (*i.e* gates G, reporters R and possibly inputs I_A, I_B are present initially). We prove that, for all initial states $q_0 \in Q_0$ and terminal states $q \in Q_{Inv}(q_0)$,

$$P_{AND}(q_0, q) \leftrightarrow q(O) = \min\{q_0(I_A), q_0(I_B)\} \tag{3}$$

holds, as long as $q_0(R) \geq q_0(I_A), q_0(R) \geq q_0(I_B), q_0(G) \geq q_0(I_A)$ and $q_0(G) \geq q_0(I_B)$. The logical behavior of the AND gate is formalized using a threshold θ where a signal represents the logical "true" if and only if the number of molecules is greater than θ *i.e.* $[q(O) > \theta] \leftrightarrow [q_0(I_A) > \theta] \wedge [q_0(I_B) > \theta]$, which implements the desired logical operation. This behavior follows directly from Eqn. (3) and was also verified using our approach for arbitrary values of θ.

OR Gate. The desired behavior of the OR gate (Fig. 6) is defined similarly to the AND gate described above. We define the set $Q_0 = \{q \in Q_{Inv} \mid q(s) > 0$ if $s \in \{R, G_A, G_B\}$ and $q(s) = 0$ if $s \notin \{R, G_A, G_B, I_A, I_B\}\}$ (*i.e* gates G_A, G_B, reporters R and possibly inputs I_A, I_B are present initially). We prove that, for all initial states $q_0 \in Q_0$ and terminal states $q \in Q_{Inv}(q_0)$,

$$P_{OR}(q_0, q) \leftrightarrow q(O) = q_0(I_A) + q_0(I_B) \tag{4}$$

holds, as long as $q_0(G_A) \geq q_0(I_A)$, $q_0(G_B) \geq q_0(I_B)$ and $q_0(R) \geq q_0(I_A)+q_0(I_B)$. As before, the logical behavior of the OR gate is formalized through a threshold θ but here, a signal represents the logical "true" only if the number of molecules is greater than θ *i.e* $[q_0(I_A) > \theta] \vee [q_0(I_B) > \theta] \rightarrow [q(O) > \theta]$. This behavior follows from Eqn. (4) and was also verified using our approach for arbitrary values of θ.

To avoid issues with the composition of multiple OR gates, the logical "false" is left undefined for this component, which is sufficient for the implementation of the square root circuit discussed next, where a dual-rail signal encoding is used.

Full Square Root Circuit. The square root circuit takes an input between 0 and 15 represented as a 4-bit binary number encoded using the concentrations of 8 input chemical species ($STRAND_0, \ldots, STRAND_7$) in dual-rail logic (see [2,17] for details). The circuit computes the largest integer smaller than or equal to the square root of the input and represents this 2-bit output using the concentrations of 4 species (M_0, L_0, M_1, L_1). Following the design from [17], the circuit is separated into a number of logical blocks (composed of the AND, OR, and FANOUT gates studied above), each of which computes a separate part of the output. Here, we study three different implementations of this circuit inspired by the design from [2]. For one (referred to as $SQRT_1$), distinct DNA domains are used to prevent crosstalk between the logical blocks. However, this increases the total number of domains required, which potentially increases the cost of circuit construction. Motivated by this, we explore two simplified designs where domains are shared between logical blocks and either the original FANOUT gate design from Fig. 2 ($SQRT_2$) or the modified one from Fig. 4 ($SQRT_3$) is used. All three circuits are designed to implement the mathematical operation

$$P_{SQRT}(q_0, q) \leftrightarrow O(q) = \lfloor \sqrt{I(q_0)} \rfloor)$$

where $I(q_0) \in \{0, \ldots, 15\}$ for $q_0 \in Q_0$ and $O(q) \in \{0, \ldots, 3\}$ for $q \in Q$ denote the input and output of a circuit, specific to each design.

For $SQRT_1$, we assume that N copies of each functional block are operating in parallel. We use the existing Visual DSD model from [2] which defines the inputs I and outputs O to capture the requirement that each circuit copy computes the correct output independently (*e.g.* $O(q) = 0 \leftrightarrow [q(M_0) = N] \wedge [q(M_1) = 0] \wedge [q(L_0) = N] \wedge [q(L_1) = 0]$ and $O(q) = 3 \leftrightarrow [q(M_0) = 0] \wedge [q(M_1) = N] \wedge [q(L_0) = 0] \wedge [q(L_1) = N]$). The strategy from Sec. 3.1 allows us to prove that $SQRT_1$ implements this behavior correctly for $N = \{1, 10^2, 10^3, 10^6\}$. Note that this circuit is distinct from the original, localized setup from [2], where only a single copy of the circuit is considered in isolation.

To obtain requirements for a population-based design that are independent of the precise numbers of molecules used as inputs, we define thresholds θ_I and θ_O where $\theta_O \leq \theta_I$. An input bit is set to true by including more than θ_I molecules of the corresponding species which defines $I()$ (*e.g.* $I(q_0) = 0 \leftrightarrow [STRAND_0 > \theta_I] \wedge [STRAND_2 > \theta_I] \wedge [STRAND_4 > \theta_I] \wedge [STRAND_6 > \theta_I]$). Similarly, an output bit is considered true if θ_O or more molecules of an output species are present (*e.g.* $O(q) = 0 \leftrightarrow [q(M_0) \geq \theta_O] \wedge [q(M_1) = 0] \wedge [q(L_0) \geq \theta_O] \wedge [q(L_1) = 0]$). In practice, the numbers of molecules for each gate of a circuit cannot be set precisely at the beginning of a computation and therefore we consider thresholds G_l and G_u where, for each initial state $q_0 \in Q_0$, $G_l \leq q_0(s) \leq G_u$ for a gate $s \in S$. This defines the set of initial states Q_0 for a dual-rail input encoding (*i.e.* where $\forall q_0 \in Q_0 . q_0(STRAND_i) = 0 \vee q_0(STRAND_{i+1}) = 0$ for $i = 0, 2, 4, 6$) where no species other than gates and inputs are present initially. Finally, we

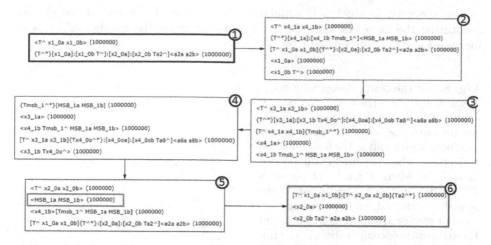

Fig. 7. For the SQRT$_3$ design with $\theta_O = 1, n = 0$ and $G_l = Gu = \theta_I = 10^6$ (*i.e.* one million copies of both the circuit and its inputs are present), initialized in state q_0 where $I(q_0) = 12$ (input is 12) a computation trace producing incorrect output was identified using the bounded model checking analysis method described in Sec. 2. In each state, only the species with nonzero abundances and participating in the reaction captured by the following transition are displayed. While part of the correct output is produced (step 5), the full output is incorrect in the terminal state (highlighted in red) and violates the dual-rail logical encoding (neither $L0$ nor $L1$ is produced). By considering multiple executions of a reaction per transition, short computation traces representing a large number of individual reactions are identified.

assume that the error with which gates are supplied $n = G_u - G_l$ is set to an arbitrary number, which simplifies the analysis but is not restrictive in practice.

We use the formulation described above to show that design SQRT$_2$ leads to erroneous behavior where, for some input combinations, no terminal states are ever reached. In other words, there exists an initial state $q_0 \in Q_0$ such that no state $q \in Q_{Inv}(q_0)$ is terminal, which violates a requirement that all computations eventually terminate.

For SQRT$_3$, we prove that the required behavior is satisfied for all input and gate settings, as long as (i) $G_l > \theta_O$, (ii) $\theta_I > 3G_u$, and (iii) $G_l > n\theta_O$. These additional constraints capture important properties of the circuit that ensure its correct operation. Constraint (i) captures the property that it is only possible to produce as much output as there are available reporter gates, (ii) ensures that enough of the input is supplied to be processed by each logical block of the circuit, and (iii) formalizes the accuracy with which the output must be measured as a function of the absolute number G_l and error n with which the circuit gates are supplied. Intuitively, to measure stronger output signal (*i.e.* where θ_O is higher) the lowest possible number of gates G_l must be increased while the error n is decreased, which might also require the addition of more input (if θ_I is higher). To confirm these requirements we used our method to find erroneous computation traces when $G_l = G_u = \theta_I$ (Fig. 7). Such behavior is also observed in stochastic simulations of a detailed models of the circuit (capturing the chemical kinetics) but becomes rare as the number of gates is increased and is easily missed (Fig. 8).

Fig. 8. For the $\mathtt{SQRT}_3$ design with $\theta_O = 1, n = 0$ and $G_l = Gu = \theta_I = 20$ (*i.e.* 20 copies of both the circuit and its inputs are present), the correct outputs are not produced for one out of the five simulated trajectories when the circuit is initialized in state q_0 where $I(q_0) = 12$ (input is 12). Output species $M1$ and $L1$ are represented by blue and red lines, respectively (output species $M0$ and $L0$ remain absent throughout these computations). The trajectory capturing the erroneous computation (reaching a terminal state at around 1500s) is highlighted.

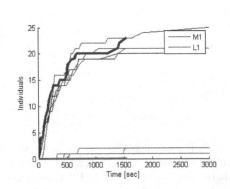

In the analysis described above, we show that if $\mathtt{SQRT}_3$ terminates, the correct output is produced. To prove that the system terminates for any choice of inputs, we employ standard Petri Net theory techniques for the computation of T-invariants [11], which reveal that, in this particular design, no cycles are possible and, therefore, termination is guaranteed.

5   Discussion

Despite the theoretical complexity of DNA strand displacement analysis problems [19], we demonstrate that an SMT-based method enables the analysis of properties of large-scale DNA computing systems and is capable of handling the largest designs currently constructed in wet labs. Although we focus on strand displacement, using the power of SMT methods and their underlying solvers to address challenging analysis questions in other DNA computing paradigms remains an interesting topic for future research. Besides improved scalability, a major advantage of the method compared to previous analysis strategies (*e.g.* [15]) is that it enables us to formalize and prove the functional correctness of systems under arbitrary inputs or large numbers of copies operating in parallel.

There are several potential uses we envision for these methods in the field. First, formalizing functionality requirements during the circuit design process and invoking the method to prove the correctness of a model is crucial in medical and industrial applications of DNA computing. For flawed designs, our method allows the identification of erroneous computation traces. Such formal "bug hunting" has become indispensable for the design of software and hardware [12] where subtle errors are hard to detect using simulation alone. Enabling a designer to explicitly state the expected functionality and related assumptions, is often sufficient to identify potential problems when reusing components in larger designs.

Furthermore, computation traces generated during analysis help identify specific inputs and conditions leading to a given behavior in the model, allowing the actual behavior of the circuit to be examined in the lab.

Second, analysis methods are useful as "compiler optimizations", for example by detecting reactions that will not be enabled for given inputs (as in our method) thereby speeding up simulation. A tighter integration with a compiler (*e.g.* Visual DSD) also benefits the analysis - our preliminary investigation suggest that, in specific cases, disabled reactions can safely be removed from the system during compilation rather than by using the more general procedure from Sec. 3.1. Incorporating analysis capabilities within DNA compilers also allows useful information to be provided to the designer and helps in understanding circuit behavior.

Despite all analysis and design efforts, our methods only allow us to gain confidence in the correctness of the available models, while the functionality of a DNA circuit is ultimately determined experimentally in the lab. Even so, models can be extended to include the additional complexities (*e.g.* unproductive reactions, leak rates, etc.) required to capture the behavior of a circuit more accurately. Furthermore, the ability to systematically analyze models has the potential to aid the construction of circuits in the lab by allowing observed experimental results that are also possible in the model (although potentially rare) to be distinguished from situations where the modeling assumptions fail.

While the analysis of large-scale, single-molecule resolution models is the focus in this paper, SMT-based methods also enable the encoding and analysis of approximations such as (non-linear) ODEs, where species concentrations are described as continuous values but important system behavior is potentially missed. In our current work, we focused specifically on a class of CRNs where computations do not depend on reaction kinetics. Recently, the class of mathematical functions computable in chemical reactions networks with arbitrary kinetics was characterized [3] and practical advantages of such systems were highlighted (*e.g.* only a set of inputs is sufficient to initiate a computation [8]). Extending the method described here to probabilistic systems to capture the additional complexity of chemical kinetics (*e.g.* through the use of stochastic SMT [10]) is an ongoing effort. For instance, in the present work, we study several DSD-circuits, inspired by the localized square root circuit from [2]. Although this design is similar to the one from [17], it does not use *seesaw* gates as a basic logical component. The seesaw gate is capable of implementing either AND- or OR-type behavior, but relies on differential binding rates between certain species to do so and, as a result, there is a low probability that the circuit will compute the incorrect output. Since chemical kinetics are not currently considered in our representation, low probability computation traces (*e.g.* where an OR-gate behaves as an AND-gate) would be identified as erroneous using our approach, without taking into account their actual probability.

In this paper we consider DSD circuits where all species and reactions are generated *a priori* (which is often the case for circuits of practical interest, but with notable exceptions [15]) and the output is measured once a state is reached where no additional reactions are possible. For more general chemical systems and other DSD circuits, this is not always the case (*e.g.* when an output signal is computed but other auxiliary reactions are still enabled). Thus, DNA computing circuits may be viewed as reactive systems that continually perform computation and react to external signals, rather than circuits that compute some output and terminate. Richer specifications (*e.g.* as captured in temporal logic) are useful for defining more general behavioral properties and are possible in the proposed framework (*e.g.* through standard encodings as in [1]). Besides capturing the functional properties discussed here, in [21] our methods proved useful for studying certain structural properties such as the presence of exposed DNA sequences in the transducer circuit designs from [15].

While termination is generally a challenging problem [23], it is possible to obtain termination proofs for many concrete models. Here, we obtain such a proof by adapting methods developed for Petri nets [11] to the particular circuit design we study in this paper. Extending this method and adapting other recent techniques to study termination in DSD circuits is a promising direction of future research. More generally, several of the properties we study are closely related to ones defined for Petri nets [11] and adapting techniques developed for their analysis is currently ongoing. Notably, the use of Petri net methods (instead of the strand-conservation strategy from [21]) to compute invariants for DSD circuits does not substitute the strengthening procedure from Sec. 3.1 for the examples we consider.

The iterative strengthening of inductive invariants (as in our strategy from Sec. 3.1) has been studied in the context of software and hardware verification, and the development of such methods for DNA circuits is being investigated within our framework. The application of such methods also provides a promising strategy for automatically uncovering important properties of circuit designs, such as the ones we defined for the square root circuit and its components. Finally, we study and prove the correctness of components of complex DNA circuits in isolation but cannot guarantee that this behavior is maintained when these components are used within larger systems - modularizing the analysis of DNA circuits is an auspicious future direction.

References

1. Biere, A., Cimatti, A., Clarke, E., Zhu, Y.: Symbolic Model Checking without BDDs. In: Cleaveland, W.R. (ed.) TACAS 1999. LNCS, vol. 1579, pp. 193–207. Springer, Heidelberg (1999)
2. Chandran, H., Gopalkrishnan, N., Phillips, A., Reif, J.: Localized hybridization circuits. In: Cardelli, L., Shih, W. (eds.) DNA 17. LNCS, vol. 6937, pp. 64–83. Springer, Heidelberg (2011)
3. Chen, H.-L., Doty, D., Soloveichik, D.: Deterministic Function Computation with Chemical Reaction Networks. In: Stefanovic, D., Turberfield, A. (eds.) DNA 18. LNCS, vol. 7433, pp. 25–42. Springer, Heidelberg (2012)
4. Condon, A., Kirkpatrick, B., Maňuch, J.: Reachability Bounds for Chemical Reaction Networks and Strand Displacement Systems. In: Stefanovic, D., Turberfield, A. (eds.) DNA 2012. LNCS, vol. 7433, pp. 43–57. Springer, Heidelberg (2012)
5. Cousot, P., Cousot, R.: Abstract interpretation: A unified lattice model for static analysis of programs by construction or approximation of fixpoints. In: POPL, pp. 238–252. ACM (1977)

6. de Moura, L.M., Bjørner, N.: Z3: An Efficient SMT Solver. In: Ramakrishnan, C.R., Rehof, J. (eds.) TACAS 2008. LNCS, vol. 4963, pp. 337–340. Springer, Heidelberg (2008)
7. Dong, Q.: A bisimulation approach to verification of molecular implementations of formal chemical reaction networks. Master's thesis, Stony Brook University (2012)
8. Doty, D., Hajiaghayi, M.: Leaderless deterministic chemical reaction networks. arXiv:1304.4519 (2013)
9. Famili, I., Palsson, B.O.: The convex basis of the left null space of the stoichiometric matrix leads to the definition of metabolically meaningful pools. Biophysical Journal 85(1), 16–26 (2003)
10. Fränzle, M., Hermanns, H., Teige, T.: Stochastic satisfiability modulo theory: A novel technique for the analysis of probabilistic hybrid systems. In: Egerstedt, M., Mishra, B. (eds.) HSCC 2008. LNCS, vol. 4981, pp. 172–186. Springer, Heidelberg (2008)
11. Heiner, M., Gilbert, D., Donaldson, R.: Petri Nets for Systems and Synthetic Biology. In: Bernardo, M., Degano, P., Zavattaro, G. (eds.) SFM 2008. LNCS, vol. 5016, pp. 215–264. Springer, Heidelberg (2008)
12. Kaivola, R., Ghughal, R., Narasimhan, N., Telfer, A., Whittemore, J., Pandav, S., Slobodová, A., Taylor, C., Frolov, V., Reeber, E., Naik, A.: Replacing Testing with Formal Verification in Intel® CoreTM i7 Processor Execution Engine Validation. In: Bouajjani, A., Maler, O. (eds.) CAV 2009. LNCS, vol. 5643, pp. 414–429. Springer, Heidelberg (2009)
13. Lakin, M.R., Parker, D., Cardelli, L., Kwiatkowska, M., Phillips, A.: Design and analysis of DNA strand displacement devices using probabilistic model checking. Journal of the Royal Society, Interface 9(72), 1470–1485 (2012)
14. Lakin, M.R., Youssef, S., Polo, F., Emmott, S., Phillips, A.: Visual DSD: A design and analysis tool for DNA strand displacement systems. Bioinformatics 27(22), 3211–3213 (2011)
15. Lakin, M.R., Phillips, A.: Modelling, simulating and verifying turing-powerful strand displacement systems. In: Cardelli, L., Shih, W. (eds.) DNA 17. LNCS, vol. 6937, pp. 130–144. Springer, Heidelberg (2011)
16. Manna, Z., Pnueli, A.: Temporal verification of reactive systems: safety, vol. 2. Springer (1995)
17. Qian, L., Winfree, E.: Scaling up digital circuit computation with DNA strand displacement cascades. Science 332(6034), 1196–1201 (2011)
18. Shin, S.W.: Compiling and verifying DNA-based chemical reaction network implementations. Master's thesis, California Institute of Technology (2012)
19. Thachuk, C., Condon, A.: Space and Energy Efficient Computation with DNA Strand Displacement Systems. In: Stefanovic, D., Turberfield, A. (eds.) DNA 18. LNCS, vol. 7433, pp. 135–149. Springer, Heidelberg (2012)
20. Wintersteiger, C.M., Hamadi, Y., de Moura, L.: Efficiently solving quantified bit-vector formulas. Formal Methods in System Design 42(1), 3–23 (2013)
21. Yordanov, B., Wintersteiger, C.M., Hamadi, Y., Kugler, H.: SMT-based analysis of biological computation. In: Brat, G., Rungta, N., Venet, A. (eds.) NFM 2013. LNCS, vol. 7871, pp. 78–92. Springer, Heidelberg (2013)
22. Z34Bio at rise4fun Software Engineering Tools from Microsoft Research (2013), http://rise4fun.com/z34biology
23. Zavattaro, G., Cardelli, L.: Termination problems in chemical kinetics. In: van Breugel, F., Chechik, M. (eds.) CONCUR 2008. LNCS, vol. 5201, pp. 477–491. Springer, Heidelberg (2008)
24. Zhang, D.Y., Seelig, G.: Dynamic DNA nanotechnology using strand-displacement reactions. Nat. Chem. 3(2), 103–113 (2011)

Author Index